Devonian Dawn: the Taconic Mountains, 380 million years ago. Painting by Kristen V. H. Wyckoff.

MAP KEY

AUTOCHTHONS

D Lower Devonian Helderberg and Tristates Groups of limestones, impure sandstones, clayey limestones, and shales. Occur at Mt. Ida (Stottville Quadrangle) and Becraft Mountain (Hudson South Quadrangle), northeast and south of Hudson, respectively.

M Mélange. Middle Ordovician megaconglomerate and megabreccia of many types, sizes, attitudes, and degrees of roundness in soft-sediment deformed Snake Hill Shale. Confined to the Low Taconics, principally onthe Hudson South Quadrangle. Formed during the Taconian Orogeny (Phase III).

PARAUTOCHTHONS

—consist of the "Synclinorium Sequence" of strata: Walloomsac black shale, slate, phyllite; Balmville Limestone; Wappinger Group of intermixed dolostones and limestones, orthoquartzite and, locally, "basement" gneisses.

NL New Lebanon Fault Slice—overlies the Rensselaer Plateau and Chatham Slices in the East Chatham Quadrangle, and underlies the Queechy Lake Slice in the Canaan and State Line Quadrangles; moved during the Acadian Orogeny.

CL Copake Lake Fault Slice—underlies the Gallatin Slice in the Claverack, Copake, and Ancram Quadrangles and the Kijk-uit [pr. "Cake-out"] Slice in the Hillsdale Quadrangle; moved during the Acadian Orogeny. The Copake Lake and New Lebanon Slices may be continuous beneath the Gallatin and Kijk-uit Slices.

ALLOCHTHONS

—consist of various stratigraphic units of the "Taconic Sequence" or, locally, "basement" gneisses.

L Livingston Fault Slice—gravity slide (in part) into Snake Hill mud; consists of Austin Glen graywacke and gray shale, Mt. Merino black shale and black chert, Indian River maroon and green shale, and maroon chert (Normanskill Group). One or all of these formations occur in the Ravena, Hudson South, Cementon, Clermont, and Saugerties Quadrangles; moved during the Taconian Orogeny (Phase III)—the earliest fault slice.

CM Curtis Mountain Fault Slice—overlies the Stottville Slice and underlies the Rensselaer Plateau Slice in the East Chatham Quadrangle; more extensive in the Nassau Quadrangle in Rensselaer County; consists of the Curtis Mountain Quartzite, Nassau olive-green and gray shale, Mudd Pond Quartzite, Ashley Hill Limestone. Moved during the Taconian Orogeny (Phase IV).

QL Queechy Lake Fault Slice—overlies the New Lebanon Slice in the Canaan, State Line, and Chatham Quadrangles; consists of the oldest strata in the Taconics—Austerlitz maroon and gray-green shales and thin reddish sandstones, Rensselaer Graywacke, and igneous greenstone flows. Conceivably, the disconnected backside of the Rensselaer Plateau Slice, implying initial transport during the Taconian Orogeny (Phase IV) and then piggybacking atop the New Lebanon Slice during the Acadian Orogeny.

S Stottville Fault Slice—gravity slide (in part) into Snake Hill mud and overlies the Livingston Slice in the Ravena, Kinderhook, East Chatham, Chatham, Stottville, Hudson North, Hudson South, Cementon, Claverack, Saugerties, and Clermont Quadrangles; moved during the Taconian Orogeny (Phase III). Includes most formations of the "Taconic Sequence"—the Rensselaer, Austerlitz, Elizaville, and Everett are notable exceptions. Locally includes Autochonous Austin Glen graywacke and Snake Hill shale.

RP Rensselaer Plateau Fault Slice—overlies the Curtis Mountain Slice in the East Chatham Quadrangle and forms the dominant topographic feature in Rensselaer County; consists of the Rensselaer Graywacke (the oldest strata in the Taconics) with associated basalt dikes and sills, and maroon and light green shales. Moved during the Taconian Orogeny (Phase IV).

CH Chatham Fault Slice—overlies the Stottville, Curtis Mountain, and Rensselaer Plateau Slices in the East Chatham, Chatham, Claverack, and Stottville Quadrangles; consists of Nassau olive-green and gray shales and argillites, thin-bedded sandstones, massive light-green quartzite and Germantown gray-black silty shales and quartzose limestone and conglomerate. This may be an example of Acadian faulting along an existing Taconian fault.

LOCATIONS

of significant rocks mentioned in text

IV Igneous: Upper Proterozoic metavolcanic greenstone; in Canaan and State Line Quadrangles.

B Middle Ordovician (Early Trenton) Balmville Limestone; in Canaan, State Line, East Chatham, Chatham, Hillsdale, Copake, and Clermont Quadrangles.

AG Middle Ordovician Austin Glen Graywacke (recently recognized as Austin Glen by the author) within Walloomsac black shale, slate in Ancram, Hillsdale, and Copake Quadrangles.

Horse: large carbonate rock sliver or fault breccia sandwiched along a fault.

? No bedrock exposures

THE RISE AND FALL OF THE TACONIC MOUNTAINS

THE RISE AND FALL OF THE TACONIC MOUNTAINS

A Geological History of Eastern New York

Donald W. Fisher, New York State Paleontologist Emeritus

with Stephen L. Nightingale

BLACK · DOME

Published by
Black Dome
Press Corp.
1011 Route 296
Hensonville, New York 12439
www.blackdomepress.com
Tel: (518) 734–6357

ISBN-13: 978-1-883789-52-7
ISBN-10: 1-883789-52-4

Library of Congress Cataloging-in-Publication Data

Fisher, Donald W.
The rise and fall of the Taconic mountains : a geological history of eastern New York / Donald W. Fisher; in collaboration with Stephen L. Nightingale.
p. cm.
Includes bibliographical references and index.
ISBN-13: 978-1-883789-52-7
ISBN-10: 1-883789-52-4
1. Orogeny—Taconic Range. 2. Geology—Taconic Range. I. Nightingale, Stephen L. II. Title.

QE621.5.T33F57 2006
551.7'0109747--dc22

2006034339

The colored Geologic Map of Columbia County was made possible in part through a grant from Furthermore: a program of the J. M. Kaplan Fund. Map artwork by Ron Toelke, based on data provided by Donald W. Fisher.

Cover photograph: Bash Bish Falls, by Vincent Bilotta. Undeniably the most picturesque waterfall in the High Taconics, west-facing Bish Bash Falls is located 0.2 miles east of the New York-Massachusetts border in the Bash Bish State Forest, Berkshire County. The rock is the Upper Proterozoic Everett quartzose argillite-phyllite-schist.

Design: Toelke Associates, www.toelkeassociates.com

Printed in the USA

10 9 8 7 6 5 4 3 2 1

Dedication

To my late beloved wife, Mary Elizabeth (Betty),
for her tranquility, enduring patience, and unflinching
support throughout my geological career,
and

to our sons, Wayne and Dale, who were exposed,
all too often, to my field trip absences from home,
but who came to appreciate the spectacular results
of geologic processes via our family travels in
the United States and Canada.

Contents

Did You Know That ...

- Marine waters have flooded New York State many times during geologic history?

- New York's youngest mountains (the Adirondacks) are composed of New York's oldest rocks?

- All of the Hudson Valley's bedrock formed when present-day New York was between the equator and 20° south latitude?

- Taconic sedimentary rocks were transported from east of present New York State, having originated as far as 75 miles (120 km)* away in present-day New England?

- During several crustal-modifying episodes, rocks were complexly folded and faulted "westward" as gigantic rock slices successively overriding one another like shingles on a roof?

- The resultant Alpine Taconic Mountains shed erosional debris westward, which consolidated into horizontal deltaic rocks that were later uplifted as an extensive plateau and subsequently eroded into the Catskill Mountains?

- From about 100,000 to 15,000 years ago, New York State was buried under mile-thick continental ice?

- Alternate advances and retreats of this titanic rock- and soil-bearing glacier smoothed the underlying rock-topography, created sediment features where none had previously existed, and plucked much of the residual soil from eastern Canada, Vermont, and northern New York and deposited it, as transported soil, within the Hudson Valley?

- Twelve thousand to 9,500 years ago, herds of elephants with "overcoats" (mastodons) trod the Hudson Valley in a northward migration route?

*One mile equals 1.6 kilometers.

Foreword

The beauty of eastern New York's countryside lives in art and legend, and the rolling topography that provides this setting is the direct result of a dynamic geologic past. Imagine a great mountain range, such as the Alps or the Himalayas, in eastern New York and western New England, the result of collisions between North America and lands to the east. Such a range existed during the Paleozoic Era, from roughly 540 million to 250 million years ago. Remnants of the deep roots of this range, part of which are known as the Taconic Mountains, extend from Quebec south to New York City. *The Rise and Fall of the Taconic Mountains* is an account of how these great mountains formed and were later eroded to their subdued, yet still majestic, shapes.

Millions of years ago these mountains were the source of sediment that was deposited by west-flowing streams, forming deltas over central and western New York. Today's Catskill Mountains are the eroded remnants of the rocks formed from this ancient Catskill delta and look eastward toward the Taconics, their source. The formation of the Catskill Mountains is therefore a significant part of this account. The counties of eastern New York, especially Columbia and Dutchess, and their neighboring counties in western Massachusetts are central to this book because their rock layers and topography reflect the geological history of the Catskill and Taconic mountains and the Hudson River Valley.

The author, Dr. Donald W. Fisher, former professor of geology and New York State Paleontologist, possesses a vast knowledge of New York State's geologic past and of its early investigators. As Dr. Fisher explains in his prologue, many students, teachers, and lay readers over the last four decades have asked whether a book or detailed map on the geology of Columbia or Dutchess County, or the Taconic or Catskill Mountains, is available in nontechnical language. Virtually no such references are readily accessible, and the few that can be obtained contain copious technical language that is largely beyond the scope of the lay reader. This accessible yet scholarly work will help remedy that situation.

Donald Fisher is the most knowledgeable geologist to have studied the geology of this region to date. He has lived in Columbia County for more than five decades and has studied New York State's geology for most of his life. As a teacher, researcher, and author of many scientific publications, he has imparted his knowledge and enthusiasm to professional geologists, amateur scientists, teachers, and students of all ages, and has inspired many to enter the profession of geology. Drawing on his fifty years of fieldwork, Dr. Fisher includes in this book his original and detailed tectonic map of Columbia County, illustrating the locations of now-inactive faults and other geological features of the region. While working for the New York State Geological Survey, Dr. Fisher and his colleagues' innovative cartography on the Geologic Map of New York (1962) set the standard for such maps—a standard now used on the geologic maps of many other states.

This volume is not only an exposition of the region's geology and an explanation of the scientific theories related to mountain building, it also contains a gazetteer of major geologic features and sites in the region. Fisher presents a history of the efforts of many geologists who worked in this area during the late eighteenth and early nineteenth centuries—the dawn of geological research in the New World—and provides many photographs, maps, geologic cross sections, and diagrams of the local rock units and their structure. A discussion of mineral resources is also provided because, as geologists are fond of saying, "If you can't grow it, you must mine it." These mineral resources—today principally crushed stone, cement, sand, and gravel—are an integral component of the region's economy.

As a friend and colleague of Don Fisher for more than thirty-five years, I continue to marvel at his enthusiasm for geology. His passion for the science and his love of helping his students to understand that science give life to his writing as he takes his readers along many memorable paths among the wonders of this region's geological past.

Robert H. Fakundiny, Ph.D.
State Geologist and Chief (Emeritus)
New York State Geological Survey/State Museum
October 2006

Prologue

"Science's only hope of escaping a Tower of Babel calamity is the preparation from time to time of works which summarize and which popularize the endless series of disconnected technical contributions."

Carl L. Hubbs, 1935

Purpose of this Work

Teachers, students, and other curious laypeople frequently ask me if there is a book and/or a detailed map of the geology of Columbia or Dutchess counties or the Taconic or Catskill mountains that they can understand. When my response is no, they (or my inner self) prod me to correct this omission. Many highly specialized articles dealing with eastern New York geology have been published in geological journals, and some pertain to the Taconic and Catskill mountains. Virtually all of these references are not readily accessible, however, and they contain copious technical language. It is my hope that this book will remedy this situation. Much of the following content is in response to questions that I have been asked during the last four decades. I have avoided using many technical terms, whenever possible, even though some geologists would chastise me for their omission.

My involvement with Taconic geology began while attending the universities of Buffalo and Rochester. There, the name Taconic traditionally appeared in structural geology courses as an example of folded and faulted mountains within a klippe—a 100-mile exotic rock perched atop underlying *younger* sedimentary rocks. This is the reverse of what would be expected; younger sediments, and thus younger sedimentary rocks, are normally found on top of older rock strata. A few years later, as an assistant professor of geology at Union College in Schenectady, New York, I had my initial field acquaintance with Taconic rocks when I collected a trace fossil (*Oldhamia*) from purple shale in the Rensselaer Formation north-northwest of Brainerd, Rensselaer County. (A complete discussion of rock formation nomenclature appears in the chapter "The Jargon of Geology.")

My appetite for Taconic geology was further stimulated because I have lived amid assorted Taconic rocks since 1955 and it was discomforting to know so little about one's local geology. But certainly the greatest impetus toward my involvement with Taconic geology came in the 1960s during my employment (1952–1982) with the New York State Geological Survey. At that time the primary project was the preparation of a colored map of the Empire State showing the distribution of bedrock geology units. Yngvar W. Isachsen was assigned the areas of metamorphic and igneous Precambrian rocks, and Lawrence V. Rickard was assigned the Devonian rocks; I was allotted the Cambrian, Ordovician, Silurian, and Late Precambrian sedimentary and low-grade metamorphic rocks. It quickly became obvious that the existing literature did not furnish adequate information for such a comprehensive undertaking. Few up-to-date geologic quadrangle maps were available. Most of the older geologic maps required field-checking and re-evaluation, and for many quadrangle maps, no geologic data was currently usable or even available. Consequently, an intensive field-mapping program was undertaken in collaboration with college geology faculty. Much new fieldwork and numerous conferences took place before a workable and understandable composite map could be compiled.

Within my area of responsibility, the region east of the Hudson River demanded wide-ranging fieldwork and re-assessment of previous work. In accomplishing this task, I produced detailed geologic maps in northern Washington, Columbia, and western Dutchess counties and less detailed reconnaissance geologic maps in Saratoga, Albany, Schenectady, Rensselaer, eastern Dutchess, eastern Ulster, and northern Orange counties—all relevant to Taconic geology. Finally, in 1962, the Geologic Map of New York State (N.Y. State Map and Chart Series #15) was made available to the public in atlas form. This consists of five sheets (Niagara, Finger Lakes, Hudson-Mohawk, Adirondack, Lower Hudson) and a sixth sheet (Master Legend). The scale of the separate maps is 1:250,000 (1 inch = 4 miles). They were reprinted as separate sheets in 1970 and 1995.

You will discover that this book often ventures far afield of the Taconic, Adirondack, and Catskill mountains and their environs. Geology is not constrained by political boundaries. To fully comprehend the mechanics of formation of these mountains, relevant geology elsewhere is necessarily discussed. In so doing, the roles played by older, synchronous, and younger rocks provide a clearer understanding of the nature of events that contributed to the history of the Taconic, Catskill, and Adirondack mountains.

Additionally, it is important to understand the system of organizing geologic periods. Mostly they have been partitioned into three parts. For *rocks units,* the divisions are Lower, Middle, and Upper (ex., Lower Cambrian, Middle Ordovician, Upper Devonian). For *time divisions* the corresponding names are Early, Medial, and Late (ex., Early Cambrian, Medial Ordovician, Late Devonian). Therefore, *Upper* Devonian rocks form in *Late* Devonian time.

Some Antiquated Nomenclature

Throughout the history of the study of geology, different systems of organizing rock strata have been used. In the early nineteenth century, for example, exposed rocks and strata were separated into four major divisions—Primary, Secondary, Tertiary, and Quaternary.

"Primary rocks" referred to the oldest metamorphic rocks. An example would be the rocks of the Hudson Highlands. Nineteenth-century workers would call these granites and gneisses "Primary rocks," while today we would say they formed during Precambrian time.

"Secondary rocks" referred to the next younger, sedimentary rocks that separate the older Primary and the younger Tertiary. Secondary rocks range through what we now call the Cretaceous.

Notes on Terminology

Where a technical geologic term is first mentioned, it is defined parenthetically or shown in **boldface,** meaning that it is explained in the glossary.

On several line drawings the following abbreviations have been used for the sake of brevity:

cgl.—conglomerate
ss.—sandstone
sts.—siltstone
sh.—shale
ls.—limestone
ds.—dolostone
ct.—chert
gray.—graywacke
mel.—mélange
oqtzt.—orthoquartzite
mqtzt.—metaquartzite
sl.—slate
mb.—marble
phy.—phyllite
sch.—schist
gn.—gneiss

In addition the following abbreviations will be frequently encountered throughout the text:

mya—million years ago
bya—billion years ago

"Tertiary rocks" referred to the sediments overlying the sedimentary rocks of the Secondary. These strata included those occurring in the Paleocene, Eocene, Oligocene, Miocene, and Pliocene epochs of current usage.

"Quaternary rocks" referred to the youngest sediments. This group includes the Pleistocene and Holocene epochs.

Within the last half of the twentieth century, a new division of the Cenozoic Era was introduced and became popular with many paleontologists and geologists. The period now known as the Paleogene Period incorporates the early part of the Tertiary Period (the Paleocene, Eocene, and Oligocene epochs). Similarly, the Neogene Period incorporates the later Tertiary and all of the Quaternary periods (the Miocene, Pliocene, Pleistocene, and Holocene epochs). As a result of this acceptance, the terms Tertiary and Quaternary are not used in this work. (As a side note, many geologists avoid using the term Holocene as it is abnormally brief in time—typically the last 12,000 years—compared to other Cenozoic epochs that are measured in millions of years. These geologists prefer to believe we are still in the Pleistocene Epoch.)

Finally, major divisions of geologic time (eras, periods, and epochs) are generally defined by major geologic or biologic events such as mountain-building episodes or mass extinctions. Scientists' understanding of these events, along with the technology used to measure when they occur, is constantly being refined. Therefore, the reader should assume that geologic dates presented in this work must be considered approximate and subject to future—and probable—adjustment. As an example, in this book the end of the Permian Period is given as 245 million years ago. Recent research on the mass extinction that terminated the Permian Period indicates that it may be linked to a geographically widespread volcanic eruption occurring 253 million years ago. If this new information achieves a consensus among geologists, future texts may date the end of the Permian as occurring 253 million years ago.

Which Way Is North?

Frequent reference to directions will be made throughout the text. The North American continent, however, has moved over geologic time—both rotating and translating. For example, present-day California was northwest of present-day New York 300 million years ago. For convenience we will refer to directions relative to today's cardinal point framework. Thus, if a rock formation exposed in Albany becomes thinner to the west, the reader may assume that "west" would be toward Buffalo—even though when the rock formed, that direction may not have been "west." Quotation marks are occasionally used to remind the reader and emphasize that all directions are specified as they exist today.

Within the geological "family," a cooperating corps of dedicated geo-scientists has revealed a captivating documentation (though incomplete) of the geologic history of our Earth. Their collective data permits us to establish the chronology, contemporaneity, diversity, and habits and habitats of past life. Within this multi-act drama we are able to reconstruct ancient environments, restore shapes, and relocate the positions of past lands and seas through time. By bouncing lasers off Earth-orbiting artificial satellites, we can measure the speed and direction of Earth's shifting tectonic plates. Mapping of modern ocean floors has reinforced our concept of plate tectonics—the solution to many heretofore geological puzzles. By studying the chemical composition, physical attributes, and fossil content of rocks and their structures, geologists confidently can ascertain how, where, and when mountains were born, how they grew to maturity and zenith, disintegrated, and eventually disappeared. This is the germane message of this book—to analyze the rock architecture and experience the times and life prior to, and during, the evolution of the Taconic, Adirondack, and Catskill mountains.

If I make you aware of the fascination and frustration of Taconic geology, the composition and construction of Catskill geology, and the complexity of Adirondack geology, then I will have succeeded in my mission.

Acknowledgments

This synthesis of my professional geologic research in eastern New York and the Taconic Mountains in particular would not have occurred without the support and encouragement of numerous friends and colleagues.

Special recognition goes to my friend Stephen L. Nightingale. Steve's experience as an earth science, physics, and astronomy teacher made him particularly suited to serve as a constructive critic toward the betterment of this project. His comments regarding the depth to which I explored the various facets of the book made the text more understandable for those readers lacking a specialized background in geology. He meticulously transferred my typescript and illustrations into the electronic format required for publication, and performed valuable text editing during that process. Photographs that Steve prepared for me attest to his skills in that area. In addition he accepted the difficult but important task of creating an index for the book. I am sincerely grateful for his unselfish dedication and encouragement toward the completion of this work.

The inclusion of the colored Tectonic Map of Columbia County was made possible in part through a generous grant from Furthermore, a program of the J. M. Kaplan Fund. The grant was administered by the Columbia County Historical Society, and I am especially appreciative of the support of the society's former director, Sharon Palmer. Ron Toelke and Barbara Kempler-Toelke of Toelke Associates designed the book and transferred my geologic observations and interpretations onto a base map of Columbia County, thus creating the Tectonic Map.

Many others generously provided their time ad assistance. One of my former students, Professor John M. Bird (retired), Department of Geology, Cornell University, read and discussed portions of my first draft and spurred me to continue the project. Doug Allen made improvements to the history section. Tyler Shafer, former worker at past and present cement plants, provided the information that enabled me to prepare a flow chart on the manufacture of Portland cement. Joseph Palumbo and Travis Bowman converted a number of my 35mm slides to electronic format during the initial phases of this project. Editor Eric Raetz suggested some restructuring and improvements to the text. Proofreaders were Matina Billias, Natalie Mortensen, Christl Riedman, and Ed Volmar.

The beautiful front cover photograph of Bash Bish Falls is the work of Vincent Bilotta. Kristen Wyckoff graciously granted use of an image of her painting Devonian Dawn, photographed by Jack Harmon. All other photographs, as well as drawings and diagrams are by the author unless otherwise credited.

Special thanks go to Dr. Robert H. Fakundiny, State Geologist of New York (retired), who read the manuscript and contributed a personal foreword whose kind words support the worthiness of my contribution. In addition, I would like to express my deep appreciation to John B. Skiba of the Office of Cartography and Publications, New York State Museum, for granting permission to use many of the images published in this work.

Lastly, Deborah Allen, owner and publisher of Black Dome Press, and Steve Hoare, editor, are commended for their encouragement and professional contributions in assembling an eye-appealing overview of this chronicle of billions of years of geologic change.

Devonian Dawn, 8' × 10' mural on display at the Gilboa Museum, artist Kristen V. H. Wyckoff, photograph courtesy of the Jack Harmon Agency, Stamford, NY. Artist's conception of the present Gilboa, Schoharie County, region as it appeared in Late Medial Devonian age. View is looking east at Alpine Taconic Mountains with tropical swamp in foreground. During this time the region was near the equator. Ferns and lycopods, with cypress-like bulbous bases (*Eospermatopteris, Aneurophyton, Archeosigillaria vanuxemi*) border the swamp, while a fish-amphibian transitional form is emerging from the water in the foreground.

part one

The Jargon of Geology

"Learn the ABC of science before you try to ascend to its summit. Never begin the subsequent without mastering the preceding."

Ivan Pavlov, 1936

Chapter 1

Minerals, Rocks, and Fossils

Minerals

Before delving into a more detailed discussion of the geology of eastern New York and environs, it is imperative to understand some basics. One is the difference between rocks and minerals. Let's use an analogy with something that everyone is familiar with—salads. Salads are *mixtures* of basic ingredients such as fruits, vegetables, nuts, bacon bits, dressings, etc. Similarly, rocks, with few exceptions, are *mixtures* of basic ingredients termed *minerals.* Over 4,400 mineral species have been identified as of the time of this writing; a few dozen new ones are recognized each year.

Surprisingly, a precise definition of a mineral is somewhat elusive. Minerals are generally defined as naturally occurring, solid, inorganic elements or compounds; most have a specific crystalline structure. Specimens of organic origin, such as amber, coal, pearl, and coral are therefore not properly termed minerals. Beautiful man-made crystals of silicon carbide commonly found in souvenir and rock shops are not true minerals; some mineralogists use the term "synthetic minerals" to describe them. Exceptions and borderline cases often arise. In ancient Greece, for example, by-products of the smelting of lead and silver ores were dumped into the Mediterranean Sea. After thousands of years, natural reactions with seawater have created a suite of materials that would not have formed without the man-made smelting process. Most mineralogists accept these as minerals. Another exception to our definition is the mineral mercury, an element occasionally found in the form of liquid droplets scattered on the host rock.

Minerals possess definite physical, chemical, and optical properties. Some of the more important physical properties useful for identification are crystal shape or habit, hardness, color, **cleavage,** luster, specific gravity, fracture, elasticity, flexibility, and

Fig. 1. Doubly terminated quartz crystals, referred to as "Herkimer diamonds," are found in cavities in Little Falls Dolostone of Late Cambrian age at various locations in Herkimer County. Crystal shown is approximately 1 inch (2.5 cm) in length. Photograph by Steve Nightingale.

magnetism. A mineral species has a fixed chemical composition, although slight variation is "permitted." For example, the mineral sphalerite (zinc sulfide) has the chemical formula ZnS. In some cases atoms of iron (Fe) may substitute for some of the zinc and thus we have [Zn,Fe]S; this is referred to as *ferroan* (containing iron) *sphalerite*. Quartz (silicon dioxide) is commonly a transparent or translucent white mineral (Fig. 1) having the chemical formula SiO_2; however, trace elements may be present in the mineral to affect the color, giving it a pink, purple, smoky, or yellow hue. These are commonly referred to as *rose quartz, amethyst, smoky quartz,* and *citrine,* but these varieties are not recognized as distinct mineral species.

Although professional mineralogists have an arsenal of sophisticated technological wonders and techniques to assist them (X-ray diffraction, electron microprobe analysis, and many others), a few simple tests performed by the amateur will suffice to identify most common minerals. The hardness test is described here because of its simplicity and importance.

Mohs Scale of Hardness (and an Aid to Learning It)

Hardness is the resistance that a smooth surface of a mineral offers to scratching. The hardness of a mineral is quantified using Mohs Scale of Hardness, devised by the German mineralogist Friedrich Mohs (1773–1839). Minerals are assigned a number from 1–10. The difference in hardness between two adjacent numbers is greater for the higher numbers and lesser for the lower numbers. Below is a list of the standard Mohs minerals, along with a mnemonic for remembering them:

1. **T**alc	**T**all	(softest)
2. **G**ypsum	**G**irls	
3. **C**alcite	**C**an	
4. **F**luorite	**F**lirt	
5. **A**patite	**A**nd	
6. **F**eldspar	**F**ellows	
7. **Q**uartz	**Q**uite	
8. **T**opaz	**T**all	
9. **C**orundum	**C**an	
10. **D**iamond	**D**ance	(hardest)

While professional mineralogists and serious amateurs may use a set of hardness "points" (small samples of the standard hardness-scale minerals mounted on a holder), some common items will suffice to approximate a mineral's hardness. A convenient initial test is to drag a corner of the mineral on a piece of glass. Glass has a hardness of 5½; if a mineral scratches glass it has hardness greater than 5; if it doesn't, the mineral hardness is less than 5½. Difficulty in determining whether or not the glass is scratched indicates that the mineral hardness is close to that of glass. With practice, the ease by which glass is scratched gives a very good approximation. Other convenient test items are: fingernails (2½); copper coins (3½); iron nails (4½); pocketknives (5½); and tool steel, such as a file (6½).

Rocks

Rocks are classified as igneous, sedimentary, and metamorphic. Igneous rocks solidify from molten or partly molten material, termed *magma* when below the Earth's surface or *lava* when it erupts onto the surface.

Intrusive (plutonic) igneous rocks cool slowly beneath the Earth's surface. This long-lasting molten state allows the mineral components to grow to large size (several millimeters or more), giving the rock a coarse texture. Examples of these intrusives are **granite, syenite, gabbro**, and **anorthosite**; although they all have similar textures, they differ in mineral composition. Dark-colored, relatively dense igneous rocks have a composition termed *mafic*. **Ultramafic** rocks are dense, dark-colored igneous rocks composed primarily of the minerals pyroxene and olivine. Light-colored, less dense igneous rocks are termed *felsic*. The coarsest intrusive igneous rocks are *pegmatites* (Fig. 2). Pegmatites have been quarried for quartz, feldspar, mica-group minerals, and beryl in the Adirondacks and in southeastern New York, but these quarries are inactive today.

Extrusive (volcanic) igneous rocks cool rapidly at the Earth's surface and therefore are fine in texture;

Fig. 2. Quartz pegmatite dike cutting olivine metagabbro along east side of Adirondack Northway (I-87), 2.3 miles north-northwest of Interchange 23 at Lake George Village.

individual grains can be seen only with the aid of a microscope. Examples of these are rhyolite (essentially a fast-cooling version of granite) and basalt (a fast-cooling equivalent of gabbro). Occasionally, igneous rocks are found that are fine-textured, with coarse crystals scattered within. This "contradictory" texture is termed *porphyritic,* and indicates a two-step process—slow cooling at first, followed by an eruption of the still partly molten mass to the surface to finish cooling rapidly. It is noteworthy that New York contains virtually no extrusive igneous rock exposed on its surface; however, anomalous exposures of volcanic rocks do occur at Stark's Knob near Schuylerville in Saratoga County, and The Knob near Queechy Lake in Columbia County. The best-known example of extrusive igneous rock in New York is the Palisades—a sill of basalt exposed for miles along the banks of the Hudson River in Rockland County.

Sedimentary rocks are those that form from three general processes and are grouped in the following categories: *clastic, chemical,* and *organic.*

Clastic sedimentary rocks form from the accumulation, compaction, and cementation of loose sediments into a solid mass; this process is called *lithification* (Fig. 3). The particle (clast) sizes and, in some cases, shapes determine the name we assign to the rock. A sedimentary rock containing a wide variety of sizes (pebbles, sand, etc.) cemented into a hard mass

Fig. 3. Glens Falls on the Hudson River looking upstream southwest from the bridge carrying US 9 and NY 32. Thin-bedded Glens Falls Limestone above, and massive Isle la Motte Limestone below.

(usually by calcite or quartz dissolved in groundwater flowing through the sediments) forms *conglomerate.* Conglomerate has the appearance of concrete. Likewise, sand lithifies to sandstone and clay becomes shale. Clastic sedimentary rocks are generally found in flat layers termed *strata;* because these layers form horizontally, any tilt or folding we see in them is an important clue to the geologist deciphering history. Shale commonly shows this layered structure; when it does not, it is often referred to as *mudstone* or **argillite.**

Careful study of clastic sedimentary rock texture can give valuable clues to the geological history of the areas in which they are found. For example, transport of sediments by the movement of water tends to round and sort particles by size; thus, angular unsorted clasts

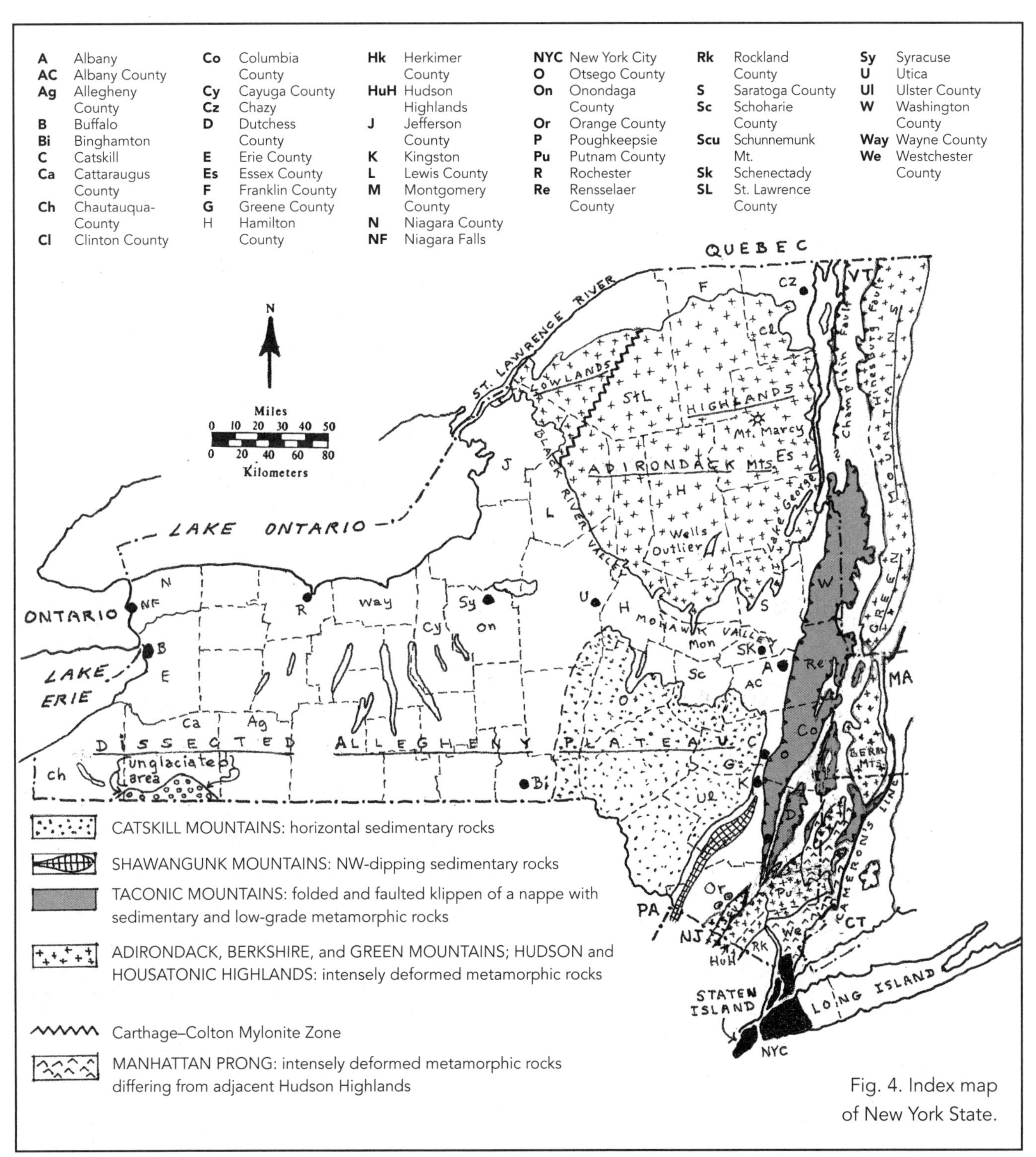

Fig. 4. Index map of New York State.

in sedimentary rock imply that the sediments were not transported very far. When sediments settle in calm water, larger particles tend to settle first, resulting in *graded bedding.* If a sedimentary rock layer is found in which the larger particles are above smaller ones, it may be inferred that this layer was overturned.

Chemical sedimentary rocks form from precipitation from solution. This occurs when the concentration of a dissolved mineral becomes too high for it to stay in solution, usually as a result of evaporation of the body of water. Microscopic particles of the dissolved material solidify (precipitate) and settle to the bottom. Over time, thick beds of salt and gypsum (both of which were economically significant resources in New York until quite recently), iron oxides, limestone, borates, and chert (either in layers or nodules) will form. Chemical sedimentary rocks are characterized by an extremely smooth texture.

Organic sedimentary rocks are formed from the lithification of the remains of animals or plants. Compacted and cemented materials such as shells and corals will form limestone (which may also form from precipitation). The burial, compaction, and distillation of plant material can form coal.

Sedimentary rocks function as time capsules because they are the only type of rock, with rare exception, in which we find fossils. New York is coated with a veneer of sedimentary rocks over the majority of its surface area (Fig. 4).

Metamorphic rocks are pre-existing rocks of any type that have been changed structurally or chemically by heat, pressure, or shearing forces within the Earth. Two general categories of metamorphic processes exist—*regional* or *contact*—and some metamorphic rocks can form from either process. Regional metamorphic rocks result from deep burial, compression, and heating of relatively large geographical areas. Under these conditions, conglomerate becomes metaconglomerate, limestone becomes marble, and sandstone alters to quartzite. Some regional metamorphic rocks become **foliated,** a term used to describe a layering or banding caused by re-crystallization of contained minerals adjusting to tremendous pressure (Fig. 5). For example, the sedimentary rock shale, under pressure and heat, will alter first to *slate;* quarries in Washington County in eastern New York produce slate for use as flagstone, flooring tiles, and roofing shingles. Greater amounts of pressure and heat on slate create **phyllite,** and phyllite can ultimately transform to **schist.** Geologists sometimes use the expression "increasing metamorphic grade" to describe the above sequence. It is important to note that in eastern New York, metamorphic grade tends to increase as we go further east—an important clue to geological events occuring here in the past.

The second process that can form metamorphic rock is termed *contact* metamorphism. This occurs when magma intrudes into the pre-existing rock. The magma "cooks" the rock it touches, resulting in a zone of altered rock surrounding the intrusion. Of interest to the mineral collector is that within the contact

Fig. 5. Tight Acadian folding of regional Taconian cleavage in Everett phyllite-schist, northeastern Dutchess County along NY 55 east of the Taconic State Parkway.

Fig. 6. Black tourmaline crystals with quartz, Bower Powers Farm, Pierrepont, St. Lawrence County. The lustrous tourmaline crystals from this locality are world-famous. Large crystal is approximately 2 inches across. Photograph by Steve Nightingale.

zone, migration of various atoms can occur between the two dissimilar rocks resulting in the growth of rare and beautiful crystal specimens. In New York, regions including St. Lawrence and Jefferson counties have a long history of producing exceptional crystal specimens of such mineral species as tourmaline (Fig. 6), diopside, and titanite.

Fossils

Fossils are evidences of past animal and plant life sealed within a tomb of sediment or sedimentary rock. On rare occasions, fossils may occur in stratified igneous rocks; for example, a lava flow may envelope an object such as a log and preserve its impression. Also,

Fig. 7. Medial Devonian brachiopods. Robust members lived in rough water, while the thin ones lived in quiet water. Photograph by Steve Nightingale.

low-grade metamorphic rocks may preserve fossils as deformed shapes. Fossils may be bones, teeth, shells, impressions, casts, molds, footprints, trails, burrows, animal appendages, exoskeletons, pollen, seeds, leaves, and petrified wood (Figs. 7, 8). Even excrement can be fossilized; these fossils are known as *coprolites.* In lieu of teeth for chewing, certain animals (for example, some dinosaurs) swallowed pebbles as a substitute for masticating food. These ingested stones are termed "stomach stones" or *gastroliths.*

Postmortem mineral replacement is fairly common in fossils. For example, in pyritized brachiopods the original calcium carbonate is replaced by pyrite. The term *silica* is frequently encountered in the context of fossilization. It is common, for example, to read that silica replaces wood or shells when they become fossilized. Silica is a *group* name for compounds that have the same chemical composition as quartz, but have varying physical properties. Opal, chalcedony, and agate are a few of the materials that fall under this category, forming *silicified* fossils. Less commonly, fossils are encountered that have a calcium phosphate composition; these fossils are termed *phosphatic.* Some strata contain few fossils, whereas others are crowded with a "fossil hash."

Paleontology is the science of animal and plant life of the past; paleozoologists study fossil animals, and paleobotanists study fossil plants. The analysis of the parade of life through

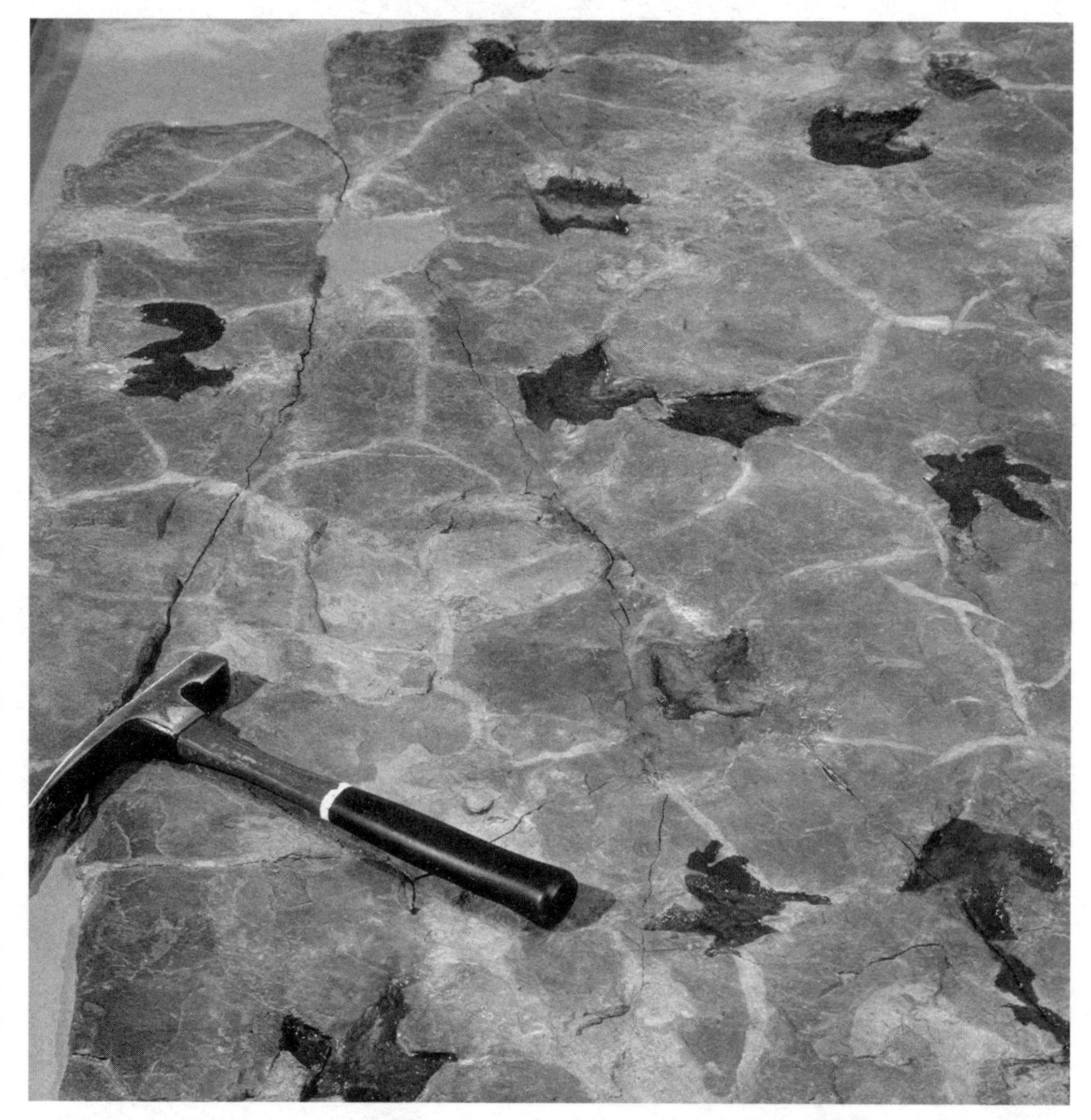

Fig. 8. Trackways of the carnivorous dinosaur *Coelophysis* in the Late Triassic–Early Jurassic Brunswick Formation (a terrestrial lakeshore shaley siltstone), approximately 200 million years old. Collected at Nyack, Rockland County.

time is based upon the features, relationships, chronology, functions, and evolution as illustrated by the fossil record. To the forensic paleontologist the numbers, kinds, and diversity of fossils in a rock provide clues to the environmental placement and habit of the once-living animal or plant. For example, research on footprints has divulged information on weight, gait, and velocity of movement of dinosaurs (Fig. 8). With careful study, inferences may be made concerning physical, chemical, and biological factors such as depth and salinity of water and biotal associates. Particularly pertinent to the reconstruction of ancient environments (paleoecology) is the proliferation of new information on the habits and habitats of past life. The discovery that some dinosaurs possessed hollow bones, together with the revelation that some dinosaurs (once living in what are now China and Argentina) were adorned with feathers, has required us to re-evaluate the traditional view that all dinosaurs were reptiles. In all probability, "dinosaurs" is an umbrella term under which some reptiles, some birds, and even some mammals have been classified. Other dinosaurs were "intermediates," while still others remain unclassified.

When unsure whether a specimen is a mineral, rock, or fossil, it may be referred to as a stone—an all-inclusive but less definitive term. If the specimen shows two or more obviously different minerals, then it is a rock. Most fossils can be recognized as such, although many sedimentary features may "mimic" true fossils; these are termed "false fossils" or "pseudo-fossils."

The Stratigraphic Name Game: Formations, Groups, and Members

Stratigraphy is the science of classifying and correlating rock units, and assembling them into a chronological sequence. A *formation* is the fundamental unit of classification. A formation is a mappable unit, and may consist of sedimentary, igneous, or metamorphic rocks, as well as unconsolidated materials such as gravel, clay, or sand. Formations may be combined into more inclusive *groups*, or subdivided into *members*.

Within a formation, composition may be homogeneous throughout; for example, it may contain only basalt. A formation may also contain repetitions of two or more kinds of rock, as with "ribbon limestone," that may contain alternating strata of limestone and shale. The thickness of a formation may range from less than a meter to several thousand meters. Formations may take from a few years to many millions of years to form.

The formal designation of a formation usually consists of two components—a *geographic name* and a descriptive *lithologic* (dominant rock type) *term*. The geographic name is derived from the *type locality*, a place where the rock is generally first identified and is clearly exposed and accessible. First letters of geographic names and rock types are capitalized; examples are *Coeymans Limestone, Rensselaer Graywacke,* and *Austerlitz Phyllite*. Occasionally rocks within a formation are heterogeneous with no single dominant type. In those cases the rock unit is designated with the locality and the word "Formation," as in *Germantown Formation*.

In the past, and occasionally even today, *faunal units* based on identification of fossils were, and sometimes are, termed "formations." This unit is far less precise, however, as time ranges of faunal units are imperfectly known within rock types, thus defeating the relatively easy application of mappable units.

Rocks Peculiar to Taconic Geology

Graywacke

Taconic geology includes some unusual rock types not found elsewhere in New York State. One is *graywacke* (the name is derived from the German word *grauwacke*, translating to "grey stone"). Graywackes with a finer texture, approaching that of sandstone, are referred to as *subgraywackes*. Graywackes consist of poorly sorted, angular to subangular clasts with a mineral composition of 50–65% quartz and a relatively large percentage of minerals derived from a metamorphic **terrane,** including biotite, muscovite, feldspar, garnet, and zircon. Because these minerals are generally derived from uplifting landmasses, their presence indicates crustal instability at the time of deposition. Whereas greywackes (sometimes termed impure sandstones) have differing compositions and environments, beach sandstones contain 85–95% quartz. There are two Taconic graywackes: the Rensselaer and the Austin Glen. Their characteristics are described as follows:

CHARACTERISTIC		RENSSELAER (Fig. 9)	AUSTIN GLEN (Fig. 10)
COLOR	*fresh fracture*	dark green-gray, green	medium bluish-gray to light gray
	weathered	light tan, dark gray to greenish-gray	light brown, orange-brown, light tan
COMPOSITION and TEXTURE		coarse-medium texture, very hard, occasionally pebble-conglomerate with salmon-pink feldspar and milk-white quartz, rare cobble-boulder congls. with light blue quartz and gneiss, clay-mineral matrix	medium-fine texture, rarely conglomeratic; if so, shows small shale pebbles, micaceous, clay-mineral matrix
BEDDING		massive, very thick-thick beds, commonly numerous secondary white-quartz veins	Upper Member: massive, very thick-thick beds Lower Member: thick-thin beds with gray shale interbeds, **crossbedded,** graded-bedded, **turbidite** features
OUTCROP HABIT		massive ledges confined to upland areas ("High Taconics")	massive to less conspicuous ledges in "Low Taconics." Abundant exposures in road and railroad cuts, waterfalls
STRATIGRAPHIC POSITION		underlies and interfingers with Austerlitz, Elizaville, and Everett Fms.	overlies Mt. Merino black shale or slate
FOSSILS		none known	graptolites in shale in Lower Member

Fig. 9. Close-up of conglomeratic phase of Rensselaer Graywacke. Photograph by Steve Nightingale.

Fig. 10. Regional cleavage (parallel to hammer handle) and bedding relationships, east limb of an anticline, Lower Austin Glen Formation, West of Highland, Ulster County. Graywacke beds are darker. Argillite and shale bedding is completely obscured by prominent cleavage.

Mélange

Another uncommon rock type restricted to the "Low Taconics" is a poorly sorted tectonic and sedimentary "blocks-in-shale" unit (Figs. 11, 12, 13). This **mega-breccia**-megaconglomerate consists of varied exotic blocks emplaced in a poorly bedded or unbedded contorted Snake Hill mudstone illustrating deformation that occurred while the sediments were still soft. These diverse transported rocks differ in age, angularity, roundness, orientation, type, and size, ranging from pea-sized pebbles to house-sized (or larger) blocks. Some outcrops display rocks of differing ages. Others contain assorted materials of the same age. Usually, various degrees of angularity, size, and rock type exist at a single exposure, although a few outcroppings show a more homogeneous composition. Such a haphazard makeup "red flags" this unit as a petrologic curiosity. Such a motley mixture has undergone a nomenclatural muddle by Taconic researchers, including breccia, megabreccia, conglomerate, megaconglomerate, mélange, **diamictite, olistostrome, wildflysch,** and chaos. Although the last-named clearly describes the jumbled makeup of the Poughkeepsie Formation, the term *mélange* has been most widely applied.

As early as 1855, Emmons recognized pebbles from different formations at Cantonment Hill (now Rysedorph Hill or the Pinnacle on modern topographic maps), east of Rensselaer. Ruedemann (1901) reported on the multi-aged fossils of a limestone conglomerate at the summit of Rysedorph Hill; he named this single large boulder the Rysedorph Hill Conglomerate (probably detached from the Middle Ordovician Balmville Limestone). But this is only an intra-conglomerate within a much larger conglomerate of the entire hill! Normally, this site would serve as a superb type locality for this mega-"plums-in-a-pudding" unit; however, because this is inaccessible private property, the exposure can no longer be examined or specimens collected. To overcome this problem, I named this significant mapping unit the "Poughkeepsie Mélange" (Fisher, 1977) with the type locality in Kaal Park, where

Fig. 11. Poughkeepsie Mélange. The large block is a conglomerate inclusion of Balmville Limestone in darker Snake Hill Shale, Ellis Avenue exposure, Cornwall Quadrangle, Orange County.

Fig. 12. Poughkeepsie Mélange with angular dolostone inclusions in darker Snake Hill Shale, Ellis Avenue exposure, Cornwall Quadrangle, Orange County.

Fig. 13. Poughkeepsie Mélange (Middle Ordovician) at type locality, east end of Mid-Hudson Bridge, Kaal Park, Poughkeepsie, Dutchess County. Austin Glen Graywacke blocks in Snake Hill Shale.

splendid public exposures are available at the eastern end of the Mid-Hudson Bridge (Fig. 13). Here, virtually all of the blocks, some quite huge, are Austin Glen Graywacke. Other reference sections occur along the railroad in and north of Poughkeepsie. The mélange is widespread in the Hoosick Falls area of Rensselaer County as the Whipstock Breccia (Potter, 1972) and is also found in the western portions of Columbia and Dutchess counties. The Whipstock Breccia may be equivalent to the Forbes Hill Conglomerate in Washington County (Fisher, 1984) and Rutland County, Vermont (Zen, 1961). It is always associated with "Emmons" fault zone and derived from the Livingston and Stottville thrust slices (Plate 1). Some instructional exposures of the mélange are at Exit 8 east from I-90 along NY 43 toward Defreestville on the south side of the highway; this exposure shows a great variety of rocks with differing orientations, all in soft-sediment deformed shale. Others are along the east side of NY 9J one-half mile south of Rensselaer, along the unnamed stream east of NY 9J just south of the Columbia-Rensselaer County line, along the Moordener Kill north of Castleton beginning at the site of the former Fort Orange Paper Company, and along Columbia County 10 south of Mt. Tom at Burden.

The mélange is sporadically found west of the Hudson River. In the early 1970s during one very dry summer, I had the good fortune of examining the exposed bed of the Mohawk River at Cohoes on the border of the Albany-Saratoga County line. East of the falls, the Poughkeepsie Mélange (in the upper Snake Hill Formation) was widely exposed over many acres, superbly displaying a proliferation of pebble- to house-sized rocks of various types, all in a shale matrix showing soft-sediment deformation. Because of the magnitude of this instructive exposure, it would have made an excellent type locality, but it is water-covered most of the time.

The rocks included within the Poughkeepsie Mélange range from the Late Proterozoic (Hadrynian) to Medial Ordovician (Mohawkian Stage)—the same range as the formations in the Livingston and Stottville thrust slices. The Snake Hill mudstone matrix yields graptolites of the *Corynoides americanus* Zone, one graptolite zone younger than the *Diplograptus multidens* Zone represented in the Austin Glen Graywacke of the Taconic Sequence—the youngest blocks in the mélange (Fig. 14). Ruedemann (1901) and later workers have reported shell fossils of Early Cambrian (Nassau), Late Cambrian (Germantown), Early Ordovician (Wappinger), Early Medial Ordovician (Chazyan), and Late Medial Ordovician (Black River, Early Trenton) ages in limestone pebbles, cobbles, and boulders (Balmville Formation) in the mélange.

In summary, field data indicate that the mélange originated when small to large detached pieces from uplifted and "westwardly" moving rock bodies spalled off and plummeted, via gravity, into a deepening oceanic basin (Fig. 15). This "north-south" elongated sea, hosting floating graptolites, was densely muddied by clay eroding from the shale-dominant terrane of the rising Taconic Mountains. Because of the erratic dumping of exotic rocks into mud (this mud was to eventually lithify to become the Snake Hill Formation), the thickness of the mélange is random, ranging from zero, where no gravity dumping took place, to about 1,500 feet, where extreme buildup occurred. The variety of ages and rock types in the Poughkeepsie Mélange argues for a derivation from an exposed elevated terrane of sedimentary rocks of differing compositions and ages. Roundness and angularity disparities denote greater or lesser distances of debris transportation.

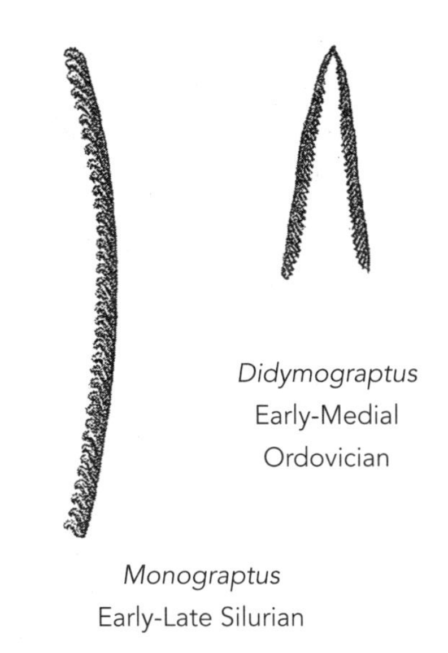

Fig. 14. Silurian and Ordovician graptolites. Printed with permission of the New York State Museum, Albany, NY.

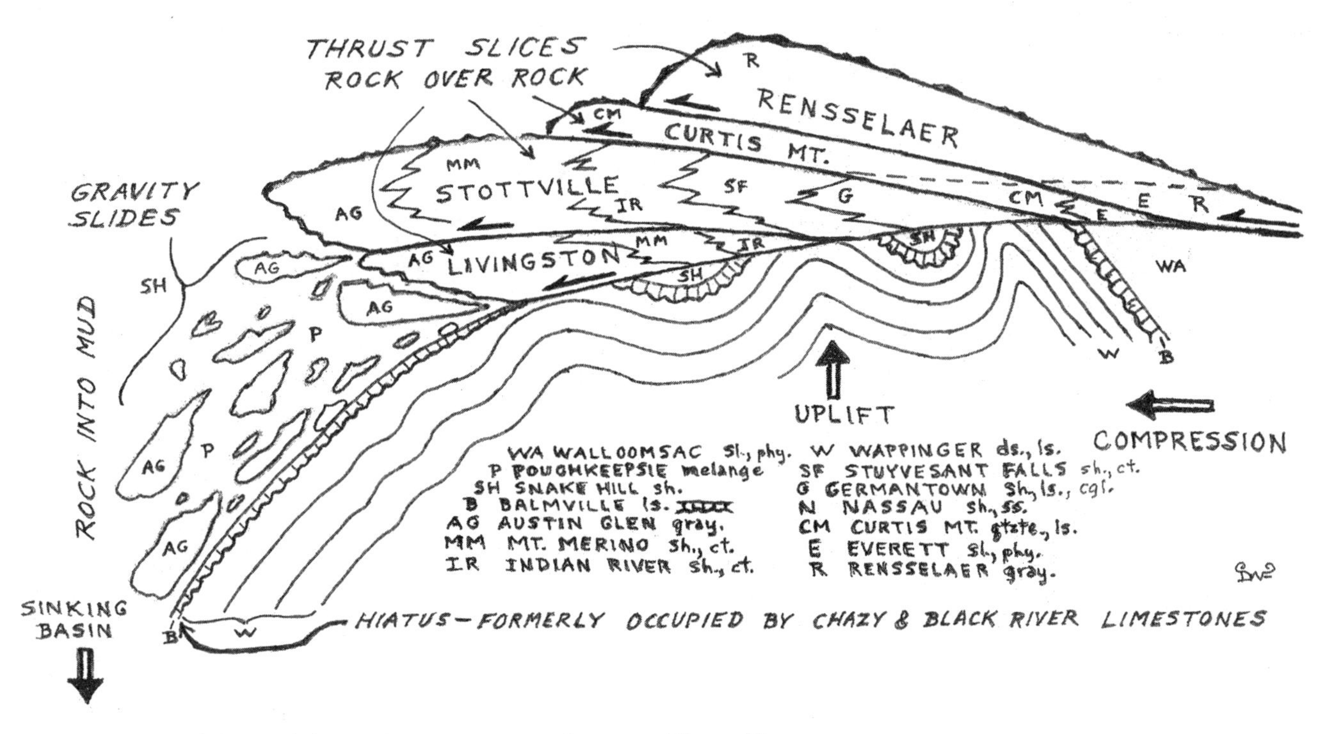

Fig. 15. Gravity slides and thrust slices—Taconian Orogeny (Phase III).

There is the potential for confusion of mélange with *tectonic fault breccias,* which also may possess heterogeneous rock makeup. Fault breccias develop where the base of overriding rock surfaces grind against overridden adjacent rock surfaces. They may be found along normal, reverse, or strike-slip faults. The fault breccias depicted on the tectonic map (Plate 1) are agglomerations of torn-up and bulldozed rock that were created when large rock slices moved over underlying rock along low-angle thrust faults. Like mélange, fault breccias display patchy distribution.

Limestone with Silicified Fossils

A singular exposure in Livingston Township on the Clermont Quadrangle deserves mention because of its uniqueness in Taconic geology. The site is a westwardly facing knoll east of the Manorton-Elizaville Road, 1.7 miles north of Elizaville. At this location is a medium gray, east-dipping (45°–60°) conglomeratic and lumpy-bedded medium-textured limestone. Silicified fossils are common and include **crinoid** columnals and **bryozoan** fragments (Fig. 16); **brachiopods** predominate and *Dalmanella, Rafinesquina, Plectambonites,* and *Triplesia* have been seen. Physical and faunal composition agrees with a Balmville Limestone designation.

At least 500 feet of exposure occurs in a west-to-east direction, but this would be abnormally thick for the Balmville Limestone in Columbia County. Undoubtedly, the section has been structurally thickened by tight folding (and perhaps faulting) giving a false thickness for the Balmville in this area. At the north edge of the knoll, there is a northwest-trending fault showing milky quartz mineralization. Whether this limestone exposure is a large limestone block surrounded by Austin Glen Graywacke exposures within a mélange, or is a limestone fault sliver left by the overriding Gallatin Slice now exposed to the east, is not clear.

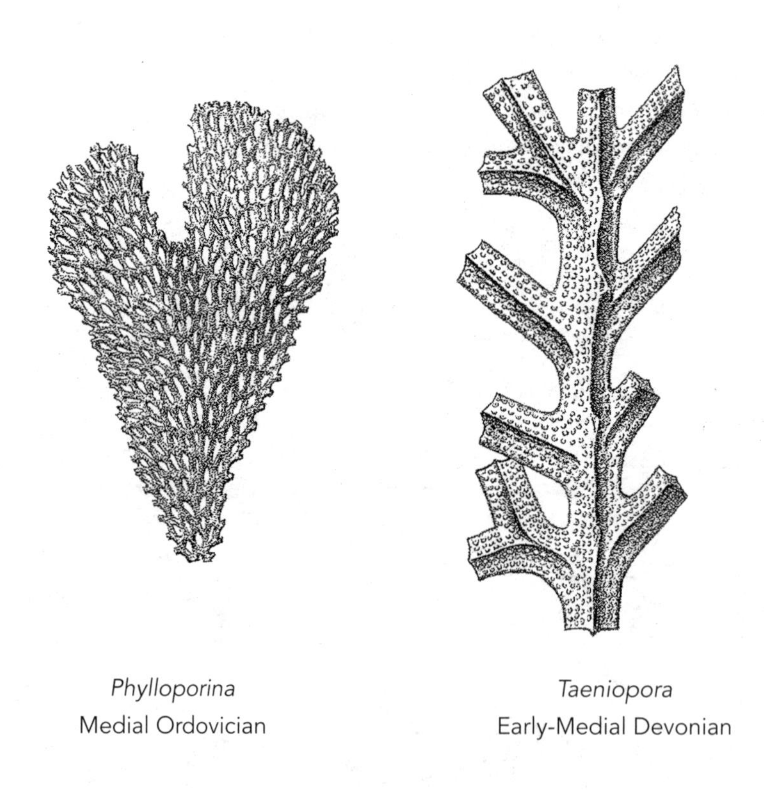

Phylloporina
Medial Ordovician

Taeniopora
Early-Medial Devonian

Fig. 16. Colonial bryozoans. Printed with permission of the New York State Museum, Albany, NY.

Chapter 2

Tectonic Plates

"The crust of our earth is a great cemetery, where the rocks are tombstones on which the buried dead have written their own epitaphs."

Louis Jean Agassiz (1807–1873)

We are all adrift on a crustal mosaic of tectonic plates, each plate "floating" in different directions at different rates of motion. These plates move on currents within the uppermost mantle in a zone termed the *aesthenosphere* (Fig. 17). In this region the pliable, taffy-like rock flows slowly because of convection driven by the high temperatures in the Earth's deeper interior. During the Earth's history, tectonic plates have been destroyed, reshaped, relocated, recycled, or rejoined.

Plate margins are sites of turmoil and thus exhibit frequent and often intense earthquake and volcanic activity. Earthquakes and volcanoes are safety valves that permit sub-crustal stresses to be released at the surface. However, earthquakes and volcanoes do not always occur on plate margins. Kilauea, on the big island of Hawaii, for example, is one of the most active volcanoes in the world, yet it is near the *middle* of the Pacific Plate. Current Hawaiian active volcanoes are passing over a hot spot in a relatively thin part of the Earth's crust. The other Hawaiian Islands, such as Maui with its spectacular Haleakala Crater, have already passed over this "pressure cooker," thus their volcanoes are now dormant. The geysers and hot springs in Yellowstone National Park occur above a hot spot near the surface of the North American Plate—perhaps a northern sub-crustal extension of the Mid-Pacific Ridge.

Journey to the Center of the Earth

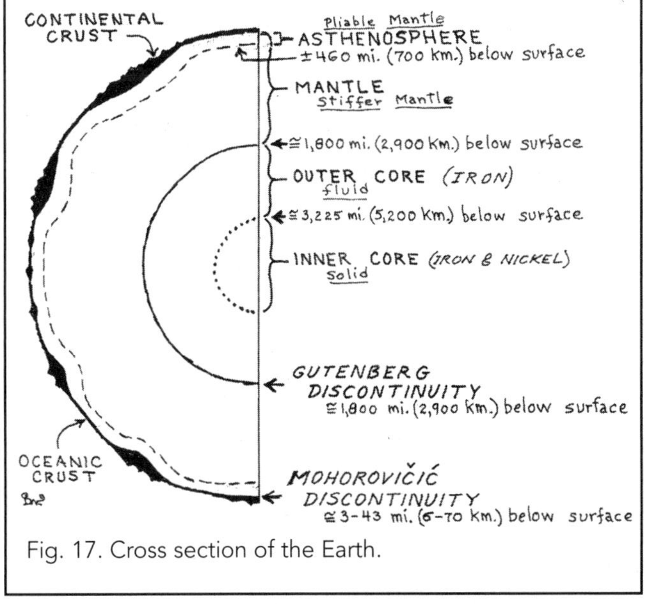

Fig. 17. Cross section of the Earth.

To better comprehend the mechanisms that drive plate tectonics, we need to understand the internal structure of the Earth. A cross section of the Earth may be compared to the makeup of an egg in which the crust is the shell, the mantle is the albumen (the "white"), and the core is the yolk (Fig. 17). The Earth's shell is divided into two distinct types of crustal plates. *Oceanic* crust consists of mafic extrusive, fine-textured volcanic rock. *Continental* crust consists of felsic intrusive, coarse-textured rock. Oceanic crust averages 3–4 miles thick with an average density of 3.0 grams/cm^3; continental crust is 25–40 miles thick with an average density of 2.7 grams/cm^3.

At *convergent* plate boundaries, tectonic plates are colliding with and moving beneath or above one another. At *divergent* boundaries, plates are separating.

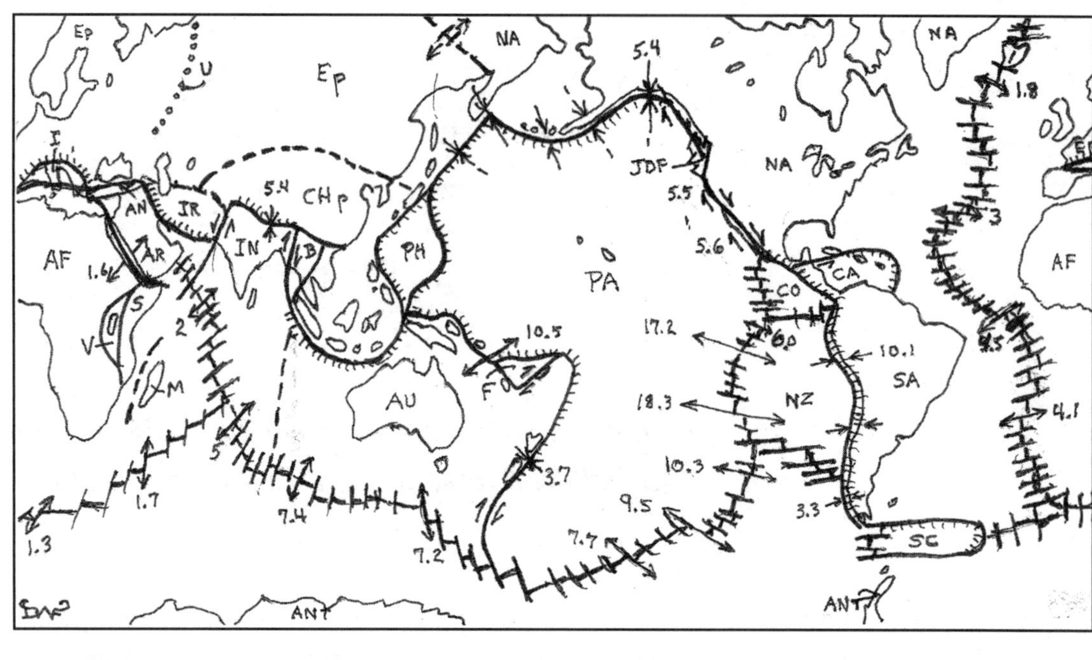

Fig. 18. Tectonic plates of the Earth's crust (compiled from various sources).

Converging plate boundaries; hachures on upper plate

Diverging plate boundaries; with *en echelon* faults. Depict sub-oceanic ridges of freshly formed igneous rock

Transform (sideswiping) boundaries; half-arrows show relative motion

Indefinite boundary; location is approximate and type undetermined

7 3.5 **Direction and rate of movement** (centimeters/year)

AF	Africa	**IR**	Iran
AN	Anatolia	**J**	Juan de Fuca
ANT	Antarctica	**M**	Madagascar
AR	Arabia	**NA**	North America
AU	Australia	**NZ**	Nazca
B	Burma	**PA**	Pacific
CA	Caribbean	**PH**	Phillipine
CHp	China plates	**S**	Somalia
CO	Cocos	**SA**	South America
Ep	Eurasian plates	**SC**	Scotia
F	Fiji	**U**	Ural ophiolite belt
I	Italy	**V**	Victoria
IN	India		

Other plates are sideswiping one another at *transform* plate boundaries. Land portions of tectonic plates are continents and islands; marine water-covered portions are oceans, seas, straits, and bays. Some plates are virtually water-covered and lack continents (Fig. 18); others are dominantly land. Still others have about equal land and water coverage. The chronicle of Taconic and Catskill geology involves the relative movements of the African, Eurasian, and the North American Plates and the consequences of their collisions and separations.

When plates converge, three types of compressional collisions are possible: oceanic-oceanic, oceanic-continental, and continental-continental (Fig. 19). Folds and thrust faults result, creating deformed mountains (like the Taconics) or island arcs (like the Japanese islands). Deep oceanic trenches develop where oceanic plates collide with continental plates; the oceanic plate "dives under" the continental plate because of its greater density, and drags a small lip of crust downward.

When plates diverge, normal faults, rifts, and block mountains form on land; mid-oceanic ridges form on ocean floors. Simultaneously, extrusive volcanic material spews forth on land as blocky lava (termed *aa*, pronounced "ah-ah") or ropy lava (*pahoehoe*, pronounced "pay-hoy-hoy"). Beneath water, the extruded magma forms *pillow lava*.

Evidence that continents move can be traced back to the time of the earliest mapmakers. A strik-

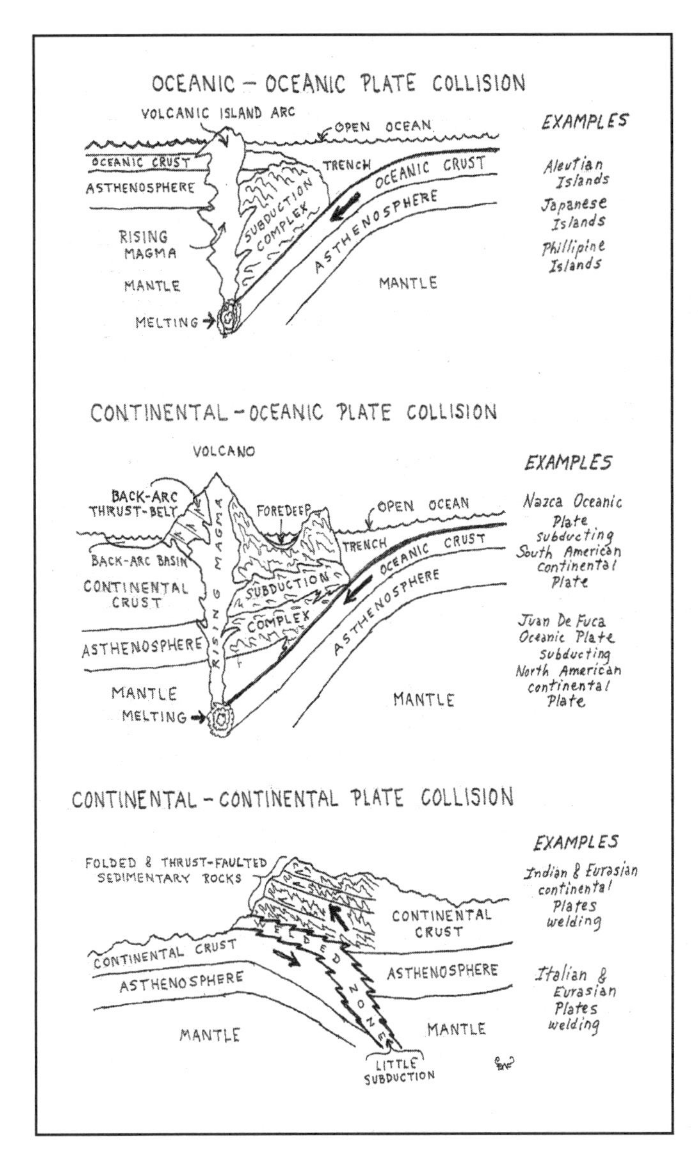

Fig. 19. Types of convergent plate boundaries.

ing and obvious "jigsaw puzzle fit" between the present coastlines of North America, South America, and Africa was—and is—readily apparent. As early as 1834, in his *Principles of Geology,* the eminent British geologist Sir Charles Lyell was using the concept of moving continents in his effort to explain the existence of tropical fossil-bearing rocks in the present polar regions, and evidence of past continental glaciers in present tropical regions. His views received scant notice. In his book, *The Face of the Earth* (1885), Austrian geologist Edward Suess coined the term "Gondwanaland" (or Gondwana) for the presumed supercontinent that comprised current-day Australia, India, South America, and South Africa. All yielded the same plant fossils (*Glossopteris* flora) of Late Permian age and which lay above Early Permian glacial **tillites**; none of the flora or glacial deposits occurred in the northern continents, termed "Laurasia" and "Eurasia." In 1908 a North American geologist, Frank B. Taylor, suggested that the submarine ridge discovered by the Challenger Expeditions of 1872–1876 might represent the site along which an ancient continent ruptured to form the present Atlantic Ocean. It was to be another fifty-three years before the implications of his speculation—that the sea floor was spreading—would be appreciated and accepted.

Alfred Wegener, a German meteorologist, set forth opinions on continental movement in his book *The Origin of Continents and Oceans* (1915). Within this work he coined the supercontinent *Pangea*. His views were received unenthusiastically by geologists, largely because Wegener's views were on a topic outside of his training. South African geologist Alexander DuToit produced *Our Wandering Continents* (1937), in which he marshalled geologic and paleontologic data supporting continental fusion and separation. For example, he noted that the Permian freshwater reptile *Mesosaurus* (Fig. 20) occurred in strata of identical ages in both eastern Brazil and western South Africa, an observation easily explained if those land areas were joined during Permian time. As a companion supercontinent to Gondwana, DuToit named the northern large land mass "Laurasia" (sometimes referred to as Laurentia).

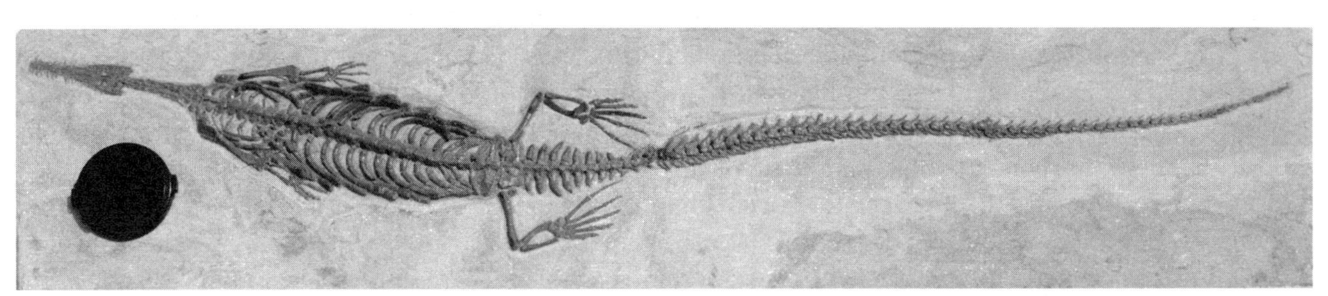

Fig. 20. Freshwater reptile (*Mesosaurus braziliensis*), Irati Formation, Permian age, Sao Paulo, Brazil. Specimen in author's collection. (Camera lens cap for scale.) Photograph by Steve Nightingale.

Although the concept of drifting continents then was accepted by an overwhelming percentage of European geologists, North American geologists were reluctant to identify with this novel theory. It was not because they were averse to new ideas, but rather that the proposed mechanisms for moving segments of crust as large as continents were unconvincing and inadequately understood. Two decades passed before a growing number of American geologists clamored to join the bandwagon of plate tectonics. This was due in part to new evidence being collected by ships crossing the oceans to measure the Earth's magnetic field as recorded in the rocks on the ocean bottom. The Earth's magnetic field reverses over long periods of time, and these field reversals are recorded in igneous rocks as they solidify. It was noticed that the fields recorded across the Atlantic were exact mirror images on either side of a ridge of mountains running through the center of the ocean. Faced with this accumulating evidence, in 1962 Harry Hess of Princeton University proposed that new ocean floor is formed at the mid-ocean ridges—the ocean floor is produced by magma that rises from deeper levels, driven upward by convection much like the motions in a boiling pot of water. Hess suggested that the ocean floor moved laterally away from the ridge and, in converging toward a continental margin, slid under the less dense continental plate. As Hess formulated his hypothesis, Robert Dietz of the U.S Naval Electronics Laboratory at San Diego independently proposed a similar model with additional details and originated the expression *sea-floor spreading*. Thus, in discovering divergent sea-floor spreading, the mechanism that forces plates to converge has also been explained, with the associated patterns of earthquakes and volcanoes that characterize these *subduction zones*.

Today the ocean bottom has been thoroughly mapped. While many chains of volcanic mountains are evident, the most obvious is the Mid-Atlantic Ridge. The creation of this new oceanic crust is generally hidden from view under the sea; however, in Iceland the volcanic action has breached the surface. The speed at which the North American, Eurasian, and African plates typically separate—approximately 1 inch/year—is comparable to the rate at which a person's fingernail grows. While seemingly slow, this is fast enough to completely reconfigure the continents over millions of years.

Additionally, the plate tectonic concept has enhanced our understanding of geologic history by the recognition of *micro-tectonic* plates. Most common along margins of lands and oceans, some recognized micro-tectonic plates are Italy, Anatolia in Turkey, Madagascar, Burma, New Zealand, Novaya Zemlya in northern Siberia, and the largely submarine micro-tectonic plate southeast of South America (Drake Passage) showing only a few islands (South Orkney, South Shetland). This discovery has been applied to help us understand ancient complex rock suites that had previously defied interpretation.

One-third of present North America displays foreign terranes that have formed elsewhere and then welded onto the North American Plate. A collage of exotic Late Paleozoic, Mesozoic, and Cenozoic micro-tectonic plates, featuring an arcuate patchwork in Alaska and a linear pattern from British Columbia to southern California, borders the Pacific Ocean. On the Atlantic side, the Nova Scotia, Piedmont, and Carolina micro-plates are Late Proterozoic and Early and Medial Cambrian attachments. What has this to do with the Taconic Mountains? Is it conceivable that the several separate major thrust slices might be sporadic emplacements of narrow, linear micro-tectonic plates? (See chapter titled "The Restless Ordovician" for discussion of the slides and slices of the Taconian Orogeny.) Only future research in geochronology, paleomagnetism, and paleontology will determine whether the micro-tectonic plate solution is a viable one for eastern New York and western New England.

Chapter 3

Faults and Folds, Slides and Slices—Disorganizers of Rocks

"The laws of nature are written deep in the folds and faults of the earth."

John Joseph Lynch, 1963

Faults and Folds

When the Earth's crust is subjected to stresses, either tensional (pulling apart) or compressional (colliding together), rocks behave in a brittle fashion and fracture along surfaces of relative weakness. If these breakage planes show no parallel displacement but move apart, the vertical or nearly vertical cracks are termed *joints*. These narrow openings act as avenues for mineral-bearing ground water and repositories for mineral crystallization. Abundant parallel joints constitute a *joint set* (Fig. 21). A *joint system* exists where two or more joint sets intersect.

If, however, there is obvious parallel movement along planar fractures, they are termed *faults*. Tensional (separating) stresses produce *normal faults* (Fig. 22). Here, the hanging wall has dropped down relative to the foot wall, and the surface lengthens or expands. Compressional (closing) stresses produce *reverse faults* (Fig. 23). Here, the hanging wall has moved upward relative to the foot wall and the surface shortens or contracts. A reverse fault with a very low angle (usually less than 15°) is termed a *thrust fault*. Sometimes one fault block sideswipes the adjacent block; the resulting vertical fracture is termed a *strike-slip* fault (Fig. 24). Occasionally, different motions (vertical, horizontal, oblique) may occur along a single fault at different times. Dormant faults may later become active, for the original fracture is a potential "release valve" to accommodate later crustal stresses.

The displacement along faults is determined by measuring the amount of offset of recognizable rock units, and in some cases, man-made features such as fences and roads. Movement direction may be determined using groove-like striations that may form along the fault. These surfaces may also exhibit a degree of polish and are referred to as *slickensided*. Fault planes and crush zones between the fault blocks are commonly sites of mineral accumulations.

Fig. 21. Typical checkered horizontal bedding plane surface of intersecting joint sets. Providence Island Dolostone Member of the Fort Cassin Formation, along Mettawee River at public fishing site, 1.2 miles northeast of NY 22 and NY 40 intersection.

You have probably observed that rocks can be cracked and broken, but have you wondered how they can be *bent*? In the Taconics, folded rocks are everywhere. The rocks must have behaved plastically because of increased temperatures and pressures. Folds are produced well below the Earth's surface where rocks are in a semi-solid, jelly-like state. Later, regional uplift brought these subterranean folds to the surface, and erosion of overlying rock exposed them to view.

Folds that arch upwards are *anticlines* (Fig. 25). A related structure is a *dome*, which is an oval or circular structure that dips radially outward in all directions from its highest point. *Synclines* (Fig. 26) are folds that are depressed downward; they are essentially inverted anticlines. In the Earth's crust, folds must eventually cease to exist longitudinally. As they gradually disappear, they are said to plunge beneath the surface and are described by such terms as *plunging anticline* and *plunging syncline* (Fig. 27).

On a regional scale, a compound anticline or compound syncline, each with intermixed minor folds, is an *anticlinorium* or *synclinorium*, respectively. In cross section, folds may be symmetrical with vertical axial planes (Fig. 25) or asymmetrical with inclined axial planes (Fig. 28)—the common Taconic type. During periods of intense compression, some anticlines may be recumbent and regionally transposed so far that the original horizontal extent of the rock formation(s) is exceeded. These structures are termed *nappes*. At this stage the strata are rupturing, creating a far-traveled overturned reclining anticline (nappe) floored by a thrust fault (Fig. 15) —a situation that occurred during Phase III of the Taconian Orogeny (to be discussed in detail in the chapter "The Restless Ordovician"). If this were the only structural disorganizer to affect the Taconic Mountains, the story of the mountains' origin would be relatively simple.

In reality, various structural complications operating at different times obscure the order of the events. Earlier folds and faults were later refolded and refaulted several times. Subsequently, the disoriented strata were metamorphosed (at least twice) progressively eastward, creating an overprinting regional cleavage. This Medial and Late Ordovician Taconian Orogeny rock disorganization was repeated, approximately 50 million years later, by folding, faulting,

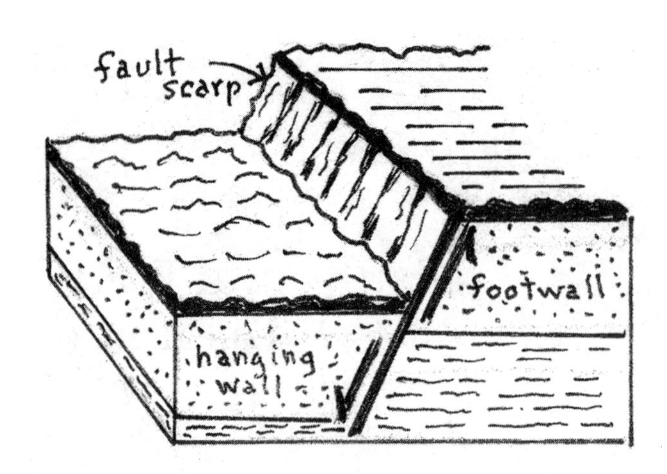

Fig. 22. Normal fault—surface expands and hanging wall drops down relative to foot wall.

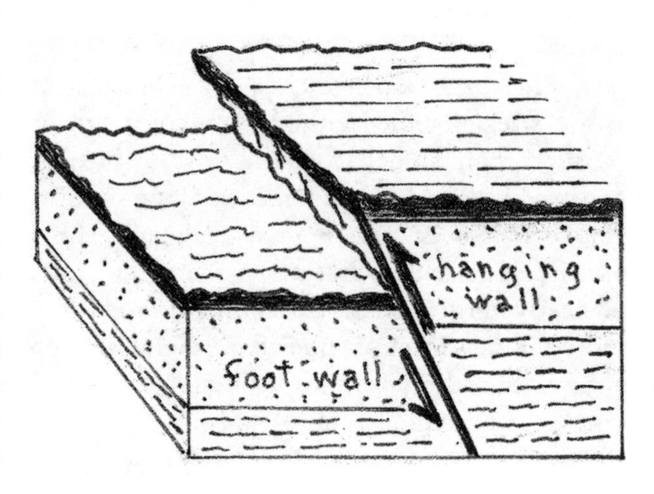

Fig. 23. Reverse fault—surface contracts and hanging wall moves up relative to foot wall.

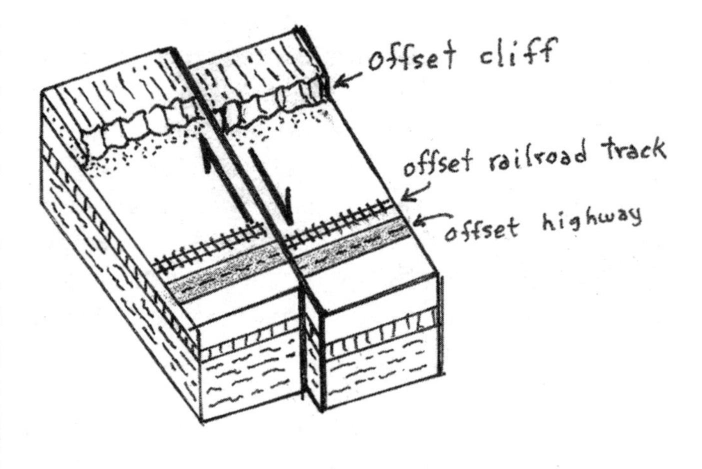

Fig. 24. Strike-slip or strike-shift fault—movement is lateral.

Fig. 25. Symmetrical Acadian upfold (anticline) with vertical axial plane in Helderbergian limestones (Becraft, Alsen, Port Ewen), along north side of NY 199 at north edge of Kingston, Ulster County.

Fig. 26. Symmetrical Acadian downfold (syncline) with vertical axial plane in Tristates Group (Esopus silty shale, below Schoharie clayey limestone), along north side of NY 199 at north edge of Kingston, Ulster County.

and regional metamorphism of the Medial and Late Devonian Acadian Orogeny. In addition, a third structural overprinting, by the Late Paleozoic Alleghenyan Orogeny, adds to the bewilderment.

Thrust Slices and Gravity Slides (Fig. 15)

A *thrust slice* (sometimes referred to as a *thrust plate* or *thrust sheet*) is a body of rock above a large-scale thrust fault whose overridden rock surface is horizontal or dipping less than 30°. Thrust slices are characterized by horizontal compression, rather than vertical displacement—as opposed to a high-angle reverse fault (dipping more than 45°), where vertical displacement is dominant.

Gravity slides result from downward movement of rock into sediment or along a rock downslope. The detachment may be sudden (avalanching) or gradual (rock creep).

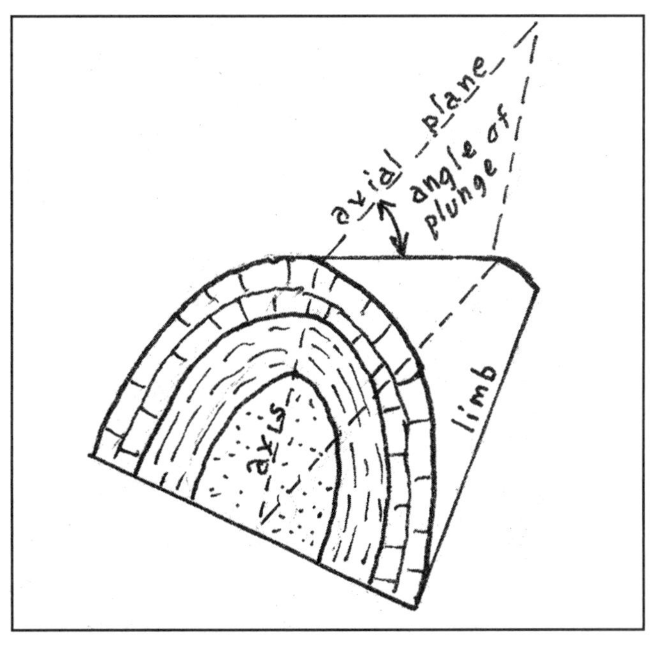

Fig. 27. Plunging anticline. (Invert diagram for plunging syncline.)

Fig. 28. Typical Taconian-type folding (westwardly overturned isoclinal folds) in ribbon limestones and interbedded black shales. Upper Hatch Hill Formation (upper Germantown), north side of U.S. 4, 4.5 miles northeast of Whitehall, Washington County.

Dip and Strike

When discussing folds, it is essential for geologists to have a means of quantifying orientations of the strata. Two quantities suffice to completely describe these orientations—*dip* and *strike*.

Dip is the angle of inclination that strata make with the horizontal, while *strike* is the compass direction of the line formed where strata intersect an imaginary horizontal plane (Fig. 29). Dip is always perpendicular to the strike. Dip is measured by an instrument called a clinometer, while the direction of the strike is determined using a magnetic compass. Both of these instruments are combined in the Brunton Compass or Pocket Transit (Fig. 30)—an indispensable tool for geologic mapping.

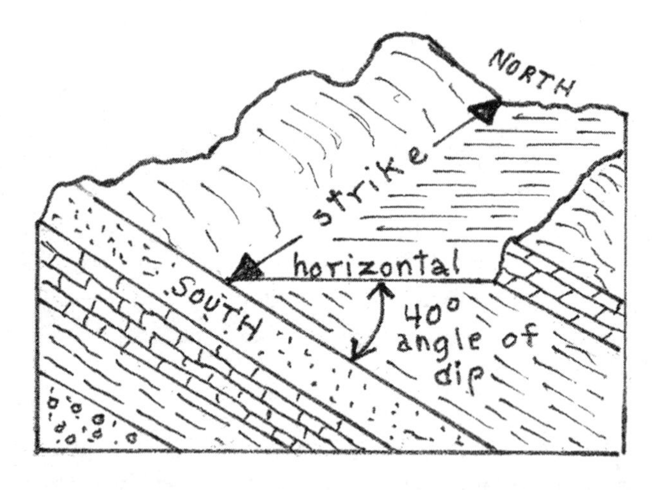

Fig. 29. Dip and strike.

Fig. 30. The Brunton Compass, posed on a section of stromatolite. Photograph by Steve Nightingale.

Chapter 4

Taxonomy: The Objectivity and Subjectivity of Classification

"The purpose of classification is not to set forth final and indisputable truths but rather to afford stepping stones towards better understanding."

L. C. Graton

Taxonomy is the science dealing with identification, description, naming, and classification of organisms. A *taxon* (pl. *taxa*) is a named group of organisms of any rank. The hierarchy of taxonomy is as follows:

KINGDOM—largest taxon; five kingdoms are customarily recognized:

Monera—bacteria, spirochetes
Fungi—fungi, molds
Plantae—red, green, and brown (kelps) algae, horsetails, mosses, grasses, ferns, shrubs, cycads, trees, flowering plants
Proctista (formerly *protista*)—**one-celled:** diatoms, sporozoans, ciliates, flagellates, foraminifera, radiolaria
Animalia—**many-celled**: sponges, corals, bryozoans, brachiopods, mollusks, worms (many phyla), arthropods, echinoderms, fishes, amphibians, reptiles, birds, mammals

PHYLUM (pl. phyla)—38 to 53, depending on the classifier

CLASS

ORDER

FAMILY

GENUS (pl. genera)

SPECIES

VARIETY

Certainly, haphazard naming of the proliferation of living and past organisms would be confusing and ineffectual. To overcome this potential chaos, Swedish naturalist Carl von Linne, alias Carolus Linnaeus (1707–1778), formulated a strategy for naming the multitude of animals and plants then known and to be discovered. His Linnaean System of Nomenclature uses form and structure of an organism as the basis for classification. Employing binomial nomenclature, the first or genus name marks a group of creatures or plants that are visibly related, such as the cats (felines) or oaks. The second (trivial) or species name applies to a restricted group within the genus. Recognizing the necessity for strict compliance to rules governing scientific nomenclature, international commissions of botanists and zoologists established the International Code of Zoological and Botanical Nomenclature; paleontologists sanction this code. This regulatory document stipulates that scientific names of organisms must be in Latin (or Latinized) and printed in italics. Furthermore, the single-word genus name must begin with a capital letter and the species name must begin with a lower case letter; it must agree grammatically with the generic name. Within a list of several species of the same genus, the full generic name need not be repeated—only the capitalized first letter. For example, in the case of felines, the panthers include *Panthera orca* (jaguar), *P. pardus* (leopard), *P. leo* (lion), and *P. tigris* (tiger). The acorn-bearers include *Quercus velutina* (black oak), *Q. rubra* (red oak), and *Q. alba* (white oak). Present-day humans sport the scientific name *Homo sapiens*. Using fundamental anatomical attributes, Linnaeus further recognized groupings of similar genera into families, families into orders, and orders into classes.

A living biological species is defined as the basic category of biological classification. It comprises related individuals that resemble one another externally and internally. They are also able to breed amongst themselves and produce similar-looking, fertile offspring. This cannot be accomplished with entities of another species. Assumed evolutionary relationships are not relevant to the botanist and zoologist; it is explicit that a biological species is objective.

Unfortunately, paleontologists do not enjoy the luxury of knowing whether similar ancient animals had the ability to breed and create fertile offspring. Neither can they use animal classifications based on soft anatomy (which is rarely fossilized) as do many present-day zoologists. Consequently, the bulk of fossil classification is based on geometry and supposed function, or comparison with presumed living relatives. Paleontological taxonomists are a dichotomous clan. By maintaining schismatic philosophical views on the relative importance of form and structural features in species and genera, classifiers become "lumpers" or "splitters." "Lumpers" are more tolerant of anatomical variation within rank, especially on the genus and species level. "Splitters" see miniscule differences to define genera or species. Consider how future paleontologists would handle a situation wherein fossils of domestic dogs required classification. Using size, tooth geometry, and bone configuration, Cairn terriers, basset hounds, poodles, bulldogs, chihuahuas, chows, dachsunds, Great Danes, St. Bernards and the horde of others, including hybrids, unquestionably would be classed as separate species—or even separate genera! But we know that all these can interbreed and bear fertile young. Modern zoologists regard all of the aforementioned as a single species—*Canis familiaris*. Understandably, paleontological species are inevitably subjective.

Chapter 5

The Enigma of Extinction

"Extinction is the rule, survival is the exception."

Carl Sagan (1934–1996)

Extinction is the annihilation of a species or higher taxon of organisms from the Earth, never to return again. It is the erasure of an entire gene pool (the inventory of genetic data that permits a species to continue to survive). At some instant in time, there remains but a sole individual of a species in futile quest of a fertile mate. When that terminal individual is unsuccessful and dies without progeny, its unique genetic code vanishes. The American mastodon (*Mammut americanum*), an "elephant with an overcoat," once flourished in New York State (Fig. 31). For the last 9,500 years the species has been nonexistent, universally obliterated. The fossil record shows that extinction is the destiny of all species.

By contrast, *extirpation* is the removal of a species or a group of organisms from one area, but the species survives elsewhere. The Barren Ground caribou (*Rangifer arcticus*) (Fig. 34) lived in present New York State about 12,000 years ago; it is absent from there now, but continues to thrive in the northern areas of Alaska and Canada. It would be incorrect to refer to the caribou as extinct in New York State; rather, they have migrated to a more nurturing environment.

The parade of life through time clearly demonstrates that specific taxa of organisms increase in numbers and diversity, reach an acme of development, then decline and become less diversified. The elephant family, for example, has been decimated to only two

Fig. 31. American Mastodon (*Mammut americanum*) browsing in its presumed habitat. Painting by Robin Rothman. Printed with permission of the New York State Museum, Albany, NY.

Fig. 33. Molar grinding surface of a Wooly Mammoth—a grazer.

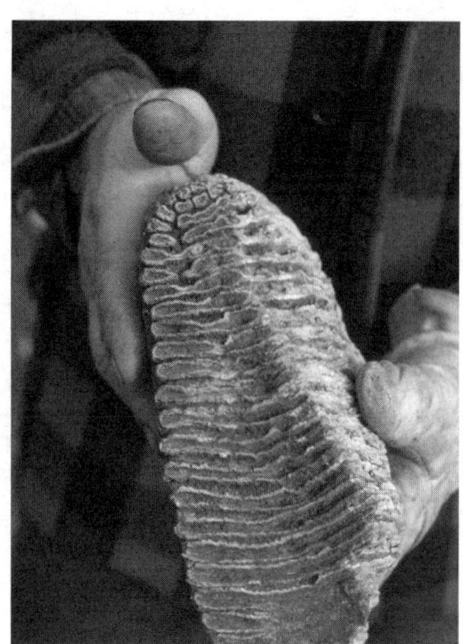

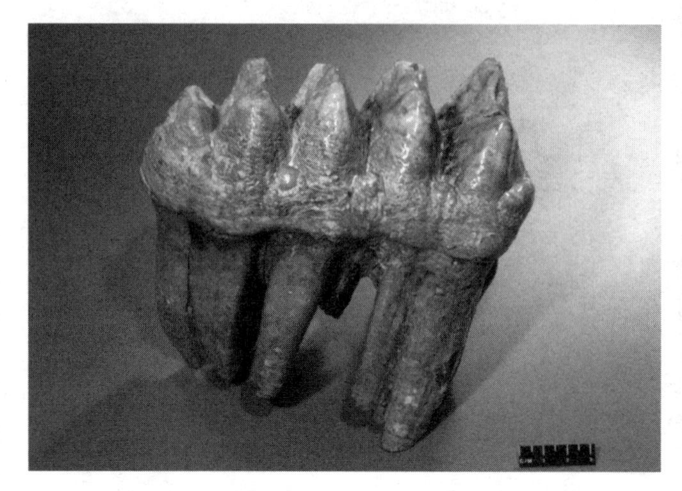

Fig. 32. Molar of American Mastodon; chewing cusps above gum line and lengthy roots below—a browser.

living species and inevitably will become extinct. Extinction of well-documented lineages may happen in less than a hundred years or may take millions of years. Plants and animals vary in their ability to tolerate, adjust to, or survive adversity. Ultimately, those species that fail to overcome calamities become extinct, while those who survive emerge into a world of opportunity rich with ecological niches left vacant by the death of their predecessors. At these moments in history, the survivors radiate into a proliferation of new forms. Every species now living owes its existence to the extinction of its forerunners.

Our best evidence indicates that mammalian genera typically have survived for less than 5 million years. Many molluscan genera have lasted for 40 million years. Genera of long-lasting identities may have existed for 100–500 million years. Do certain organisms persevere because they dwell in static, nurturing environments, enjoying longevity because of the absence of predators, or because they possess superior genes? Some biologists now believe that a major factor in the survival of a species is luck; the fossil record is rich with "body plans" that, while they seemed perfectly suited to their environment, failed to succeed as a species (Gould, 1989).

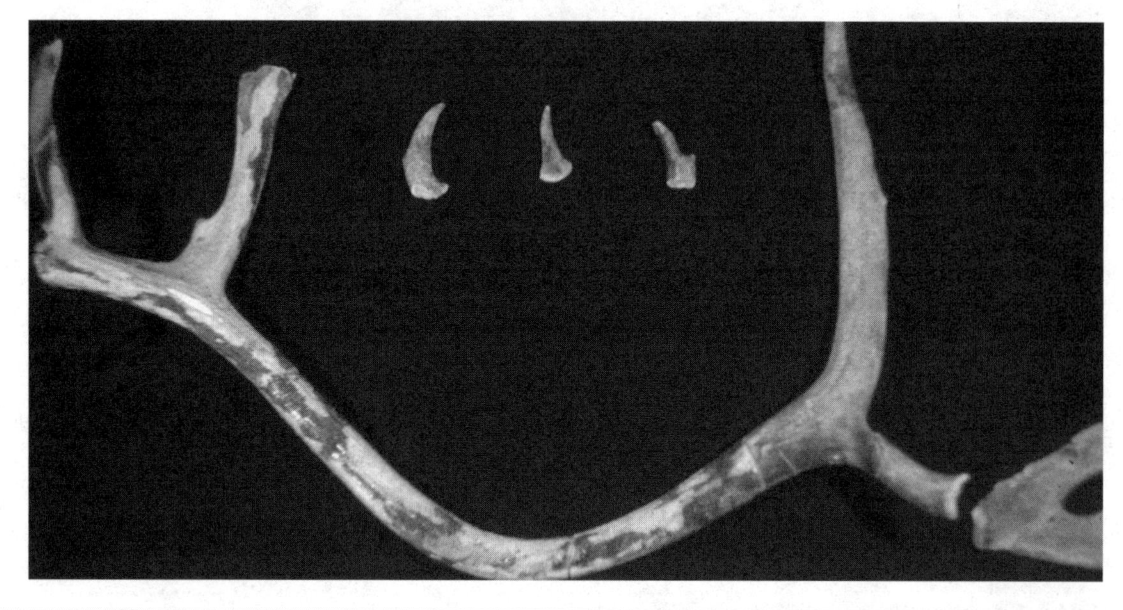

Fig. 34. Fossil antler and tines of Barren Ground Caribou (*Rangifer arcticus*), from Late Pleistocene sand and gravel pit north of Schenectady.

Mortalities on a global scale are large-scale evolutionary phenomena, but their recognition is somewhat predicated by an imperfectly known fossil record, sampling inadequacies, and inconsistent taxonomic practices. Paleontologists have documented in the fossil record five primary and some ten secondary mass exterminations occurring from Medial Cambrian to Pleistocene ages. The five primary events are as follows, with the estimated extinction percentages of the families existing at that time:

- **end of Late Ordovician** ≈ 435 mya; 22% of families disappeared
- **during Late Devonian** ≈ 362 mya; 22% of families disappeared
- **end of Permian*** ≈ 245 mya; 50% of families disappeared
- **end of Triassic** ≈ 200 mya; 20% of families disappeared
- **end of Cretaceous** ≈ 65 mya; 15% of families disappeared

**The "Mother of all Extinctions"—an estimated 57% of all families and 95% of all marine species went extinct. This extinction coincides with what is believed to be the largest volcanic eruption in the Earth's history, forming the western "Siberian Traps"—a 3 million km^3 basalt flood. In one possible scenario, this eruption initially caused dramatic cooling and glaciation in the high latitudes, quickly followed by devastating greenhouse heating, acid rain, and ozone depletion (Erwin, 1993).*

Attendant to these obliterations, the Earth's biota was severely depleted. Scattered survivors faced little rivalry for food and living space among the uncrowded living realms. This condition of non-competitive geographic isolation triggered *Adaptive Radiation*— the burst of many new species having partially inherited characteristics (hybrids) of those that outlived their doomed cohabitants. Evolutionary diversity was effectively accomplished. Ironically, extinction accelerates speciation—out of calamity comes opportunity.

Following are some possible reasons for extinction (with possible representative examples). Two or more of these possibilities may operate independently, concurrently, intermittently, progressively, or synergistically to accomplish the process of extinction.

- Failure to adapt to marked salinity changes in water (eurypterid demise?)
- Inability to adjust to drastic temperature change over prolonged periods
- Change in elemental composition of the atmosphere, such as decrease in oxygen causing asphyxiation (demise of the dinosaurs?)
- Universal poisoning of the air, food, or water by toxic elements or compounds (acid rain, selenium or sulfur poisoning)
- Dramatic and expansive lowering of sea level with resultant elimination of vast shallow water environments, forcing entire populations to collapse because of concentration of diseases and elimination of food supplies (ammonoid **cephalopod** demise)
- Annihilation by pandemic fatal diseases on land and in water (plagues)
- Lack of sufficient and/or nutritious food, causing breakdown in immune systems and ultimate starvation (demise of larger dinosaurs, ex. Sauropods)
- Destruction of a restricted food supply and the unwillingness or inability to substitute a new diet (this is the present precarious position of the Koala, which eats eucalyptus exclusively)
- Competition for living space, resulting in environmental ouster and failure to adapt to the exiled environment
- Population elimination because of excessive predation (extinction of **trilobites** by cephalopods and fishes) (Fig. 35)

- Genetic imbalance, producing faulty glandular function and defective motor coordination
- Difficulty of procreation yielding infrequent births; the death rate surpasses the birth rate
- High-dosage radioactivity, producing abnormal or stillbirths and sterility in one or both sexes
- Widespread intense volcanic eruptions, which would obliterate food supplies and/or induce asphyxia from contaminated air
- Extraterrestrial collision by meteorite* shower, asteroid, or comet. Depending on the size of the extraterrestrial object and whether it impacted on land or in water, the collision could produce a holocaust with regional or global ramifications. Hurricane-force winds and atmospheric blackout of the sun would make chlorophyll manufacture difficult or impossible. Plants could no longer propagate and would die. Expunging the Earth's flora would set up a chain reaction whereby plant-eating animals would vanish and meat-eating animals would succumb for lack of prey.

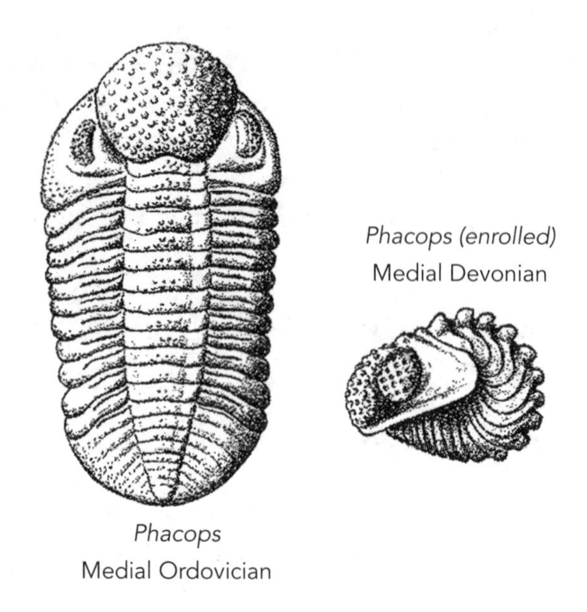

Fig. 35. Medial Devonian trilobite (*Phacops rana*), extended and enrolled. Printed with permission of the New York State Museum, Albany, NY.

*Meteoroid *is the term used to describe the rocky, metallic, or icy object while still in space; if it survives passage to the Earth's surface, it is termed a meteorite. While larger meteoroids are termed asteroids, there is no clearly defined size distinction between meteoroids and asteroids.*

The prospect of an "Armageddon asteroid" collision with Earth has received much publicity. Extraterrestrial bodies have been smashing into the Earth since its beginning. The best-known impact crater, and the first to be recognized as such, is the Barringer Meteorite Crater in Canyon Diablo near Winslow, Arizona.[1] With a diameter of .75 mile and a depth of about 750 feet, it is the most visited impact crater in the world (Fig. 36). An avid student of impact craters, the late geologist Eugene Shoemaker, calculated that the object that created the Barringer Crater about 50,000 years ago was approximately 80 feet in diameter. Its estimated weight was 63,000 tons, and it plummeted to Earth at a velocity of 10 miles/second at an angle of 30° to the horizontal. The object was almost completely vaporized during the impact, although fragments of nickel-iron are still found and are referred to as Canyon Diablo meteorites.

Fortunately for us, most meteorites are prone to disintegrate from friction with the atmosphere. The largest single meteorite known is the Hoba Meteorite in South Africa, measuring 9½ × 9 × 3½ feet and estimated to weigh 60 tons. To date, more than 150 authenticated impact structures have been identi-

Fig. 36. Barringer Impact Crater, Canyon Diablo, near Winslow, Arizona. Three-quarter mile in diameter, 750 feet deep. Photograph by D. Roddy, courtesy USGS Astrogeology Research Program, http://astrogeology.usgs.gov.

fied on the Earth's surface. Among these are some impressive circular basins marking the impact sites of former celestial invaders. At least 17 of these surface scars are between 10 and 50 miles wide. Seven craters are more than 50 miles in diameter; they are:

NAME, LOCALITY	DIAMETER (Miles)	EXPOSED	APPROXIMATE AGE (mya)
Vredefort Ring, South Africa	186	yes	2,023 ± 4
Sudbury Structure, Ontario	155	yes	1,850 ± 3
Chicxulub, Yucatan, Mexico	112	no	64.98 ± 0.05
Manicouagan Lake, Quebec	62.1	yes	214 ± 1
Popigai Basin, Siberia, Russia	62.1	yes	35 ± 5
Ancraman, South Australia	55.9	yes	> 450
Chesapeake Bay, Virginia	52.8	no	35.5 ± 0.6

Ohio State University researchers have reported a 300-mile-wide crater under the East Antarctic ice sheet. Probably caused by a comet 30 miles wide, the impact occurred about 250 mya during the Permian extinction, when most life on Earth perished.

Currently most books on dinosaurs attribute their extinction, together with that of other contemporary life, to the Chicxulub asteroid impact at the Mesozoic-Cenozoic boundary 65 million years ago. Overlooked is the fact that every major extinction event correlates with a widespread lowering of sea level. Certainly, the estimated 6-mile-wide Chicxulub asteroid shocked the Cretaceous world by depositing a layer of dust containing 30 times the amount of iridium found in the Earth's crust atop the youngest Cretaceous sediment. In addition, liquified ejecta from the impact site showered large regions with droplets of shattered, shocked and melted quartz and other silicate materials forming variously shaped **tektites.** The fossil record illustrates, however, that most dinosaur families were already extinct long before this asteroid hit.

Meanwhile, halfway around the Earth and at about the same latitude, a colossal volcanic flooding was taking place in what is now India. The result was the thick, far-flung basaltic Deccan Highlands. There, so much lava was extruded that enormous amounts of gas and ash permeated the atmosphere, effectively blocking sunlight from reaching the Earth's surface. This contaminated air shield would have wreaked havoc on plants, depriving them of chlorophyll-producing sunlight. Animals would have experienced starvation from plant-deprivation and asphyxiation from clean-air deprivation.

The one-two punch of these dual cataclysms—one from within the Earth and one from outer space—probably operated together to sound the death knell for many land animals. The last dinosaur stragglers—whose extinction had already been predetermined—were finally "down for the count."

Although the airborne pterodactyls (now considered to have been birds or mammals by some paleontologists), reptilian marine mosasaurs, and plesiosaurs vanished,[2] some hardy land animals weathered the storm of the Chicxulub–Deccan events. Surviving examples are small insect-eating mammals, crocodiles, turtles, lizards, snakes, frogs, most orders of insects, land mollusks, mosses, ferns, horsetails, cycads, ginkgoes, conifers, and flowering plants. Except for the ammonoid cephalopods, which probably disappeared for other reasons, marine invertebrates and fishes were little affected.

A couple of additional plausible extinction causes for dinosaurs deserve mention. The sediment within the Chicxulub Crater is rich in gypsum ($CaSO_4 \cdot 2H_2O$). In water this produces sulfuric acid (H_2SO_4), which produces acid rain when released into the atmosphere. This noxious by-product shows how a cosmic mishap may have altered life on Earth. Another novel extinction mechanism, proposed by Professor Keith Rigby of Notre Dame, is based on analysis of trapped air within bubbles of Cretaceous amber. Because amber is impermeable, the composition of the Cretaceous atmosphere should be preserved. Mass spectrometer analysis reveals that the Cretaceous atmosphere may have contained as much as 35% oxygen; today, our air contains 21% oxygen. If this presumed vital oxygen loss transpired as a result of the Chicxulub impact, dinosaurs may have died from oxygen deprivation.

Within the extinction arena are catastrophists in one corner and gradualists in the opposite corner. The former believe that most, if not all, exterminations are caused by sudden catastrophes such as asteroid impacts, volcanic eruptions, earthquakes and tsunamis, floods, droughts, and plagues. The gradualists believe that various causes working unilaterally or concurrently in force over long periods determine the extermination of animal and plant groups. Both views are valid and demonstrable. But the causes for all known extinctions are not yet provable. For the dinosaurs, their doom must remain a perplexing mystery.

Chapter 6

Evolution

"All naturalists who have carefully examined the arrangement of the mineral masses composing the earth's crust, and who have studied their internal structure and fossil contents, have recognized therein the signs of a great succession of former changes."

Charles Lyell, 1833

There is nothing so ubiquitous or permanent as change. Those changes that play little or no role in our busy lives or which operate so slowly as to appear ineffective usually go unnoticed. Nevertheless, geological and paleontological changes are inexorable and ongoing.

Rocks are subject to change over the course of time through the processes of weathering and erosion. Weathering is the destructive action (or group of actions) which causes earthy and rocky materials to become altered by atmospheric, chemical, and physical agents. Such action may cause the alteration of color, composition, texture, hardness, or form. Lacking transport, it is the physical disintegration and chemical decomposition of rocks that produce *in situ* sediments. Erosion begins when these sediments or rocks are moved or sculpted by wind, gravity, ice, or water, or by a combination of these erosional agents. (Running water or moving ice, by themselves, cannot physically erode bedrock. Rather, they abrade and pulverize bedrock by virtue of the volume and type of enclosed sediment or rock within them. Thus, ice, with its enclosed sand, gravel, and larger rocks, may act like sandpaper as it slowly grinds over the bedrock below. Likewise, the rolling and bouncing [traction] of sand and gravel on the bed of a stream will widen and deepen it. Wind, with its contained sand and silt particles, may also be an effective agent of erosion, especially in arid regions.)

Within this context of inexorable change, fossil animals and plants succeed themselves in an orderly and determinable sequence. This *Law of Paleobiological Succession* stipulates that each defined age within the Paleozoic, Mesozoic, and Cenozoic Eras can be identified by its diagnostic fossils. These index fossils are dependable indicators to specified intervals of time. Ideally, they were mobile families, genera, or species that were short-lived (relative to geological time), geographically widespread, relatively abundant, and easily identifiable by paleontologists with expertise in specific fossil groups. For example, an expert on trilobites can determine the precise age for Cambrian through Permian Periods—the time range for the existence of trilobites. Similarly, those who specialize in dinosaurs can accurately place their respective lifetimes within the Triassic–Jurassic–Cretaceous Periods.

Specialists in the many different groups of animals and plants complement each other in the accurate assignment of fossils and their enclosing rocks within the Geologic Time Scale. The validity of separateness of key fossils in rocks of differing ages is the domain of *biostratigraphy*. One of the pioneers of this branch of paleontology was the nineteenth-century French geologist-zoologist Alcide d'Orbigny, though the term biostratigraphy was not coined until 1904 (by Belgian paleontologist Louis Dollo). Meticulously collecting Jurassic and Cretaceous fossils from France and Germany, d'Orbigny formulated a system, still in use today, for subdividing strata into stages—the sequential pages in the stratigraphic "book of life." But the efforts of biostratigraphers only furnish a relative Geologic Time Scale. The discovery of natural radioactivity near the beginning of the twentieth century permitted the creation of a more precisely dated time scale whereby actual ages, in years, could be assigned to rock units.

Among the biological and paleobiological sciences: comparative anatomy demonstrates that evolution is possible; embryology demonstrates that evolution is probable; and paleontology (specifically, the sequential fossil record) demonstrates that evolution is a fact, not a theory.

The chronological story of fossils reveals that not all animals and plants evolved at the same rate. Some taxa changed relatively rapidly, whereas others exhibited modifications over a long duration—and a few have even resisted evolution. Interestingly, many forms of life seem to have evolved by "fits and starts"—rapidly for a short period of time, then slowly over a succeeding longer period. This concept, promoted in 1972 by Niles Eldridge and Stephen Jay Gould, has been coined *Punctuated Equilibrium* and served as a catalyst in changing the way scientists view evolution. Punctuated equilibrium explains why species have persisted either relatively unchanged or have undergone quick "spurts" of change, in contrast to *gradualism* as espoused by Charles Robert Darwin (1859, *Origin of Species*) and George Gaylord Simpson (1944, T*empo and Mode in Evolution*). Those living organisms that have withstood evolutionary changes have been popularly termed "living fossils"—an oxymoron, since living animals and plants *can't* be fossils. Though Darwin coined that term, it contradicted his view of evolution since it appeared that not all creatures evolve. Some examples of such "permanent identities" (an expression used to label a species that has shown no *discernable* evolution) are:

- **Stromatolites**, in existence for approximately 3 billion years, are seemingly changeless entities that survive today in a high salinity, brine-bath environment poisonous to most organisms
- The inarticulate brachiopod *Lingula*, virtually unchanged since the Early Cambrian (≈543–511 mya)
- Modern sharks, not unlike those from the Devonian Period (390 mya)—a relic of stagnant evolution
- *Coelacanth*, a lobefin fish, has been identified in fossils found in the Cretaceous chalk of England; living examples have been caught by fisherman off the Grand Comoro Islands near Madagascar
- The Gingko tree, with its characteristic fan-shaped leaf, is a Cretaceous leftover from its 90-million-year ancestors

Darwin's unparalleled contribution was confronted during his time with at least five major obstacles:

1. A lamentably inadequate fossil record (today our paleontological data is significantly more comprehensive).
2. A lack of mechanisms by which adaptive characteristics could be transmitted from generation to generation (Gregor Mendel's principles of genetics, and the blueprint of genetic inheritance [DNA], were yet to be discovered).
3. A hostile religious opposition loathe to acknowledge change over time.
4. Inability to communicate with a largely illiterate population.
5. Failure to convince eminent opponents in a cliquish scientific community.

The last three obstacles resulted in an "Evolutionary War" between protagonists—led by Darwin, Thomas Henry Huxley, Asa Gray, and Sir Charles Lyell—and a multitude of antagonists.

Darwin collected his profuse and extensive data while serving as the "official naturalist" during the voyages of the *H. M. S. Beagle*. Especially significant were his studies in the Galapagos Islands off the mainland of Ecuador. He placed the creation and evolution of species on a scientific footing by promoting the concept of *natural selection*. This is the natural process that results in the survival and reproductive success of groups or individuals best adjusted to their environment, and which leads to the perpetuation of genetic qualities best suited to a particular environment. Natural selection favors advantageous traits in organisms, such as acute hearing, superior eyesight, longer legs, stronger beaks, more versatile chewing apparatus, and opposable digits. If these attributes provide an advantage for survival, individuals within a given species that possess these attributes have a greater probability of procreating. These advantageous traits may then be passed on to offspring and hence, over many generations, evolutionary change occurs.

Paleontologists enthusiastically and zealously pursue the morphology, physiology, and chronology of fossils in order to test evolution as the paleobiological instrument of change.

Science and Religion—A Personal View

Often, science and religion are viewed as operating at cross-purposes, but in actuality, religion and evolution are not incompatible, for they address quite different aspects of life. Scientists are passionately curious about how, why, and when natural processes operate. Religions are concerned with dogma, faith, morality, and human well-being on Earth and in the possible hereafter. Some fundamentalist sects are openly hostile to science, but to attack science—for example, evolution—in the name of religious orthodoxy is detrimental to both science and religion.

Why do paleontologists, other scientists, and many laypersons believe in evolution? Despite claims by creationists—opponents of evolution—the entire disparate animal and plant world does not suddenly appear at the bottom of the fossil record. Opponents of evolution maintain that those taxa that have not endured today were destroyed by Noah's flood. If so, one would expect all the extinct animals and plants to be admixed in a single sedimentary formation—the one that accumulated in "Noah's Sea." Scientists have failed to discover this worldwide assemblage. On the contrary, paleontologists have definitely determined that ancient taxa lived at different times and became extinct at different times; accordingly, these multi-aged fossils occur in discrete age strata. For example, no single sedimentary mapping units contain (or will ever be found to contain) the following animal combinations, though each pair had similar habitats:

Trilobites[3] (*Early Cambrian–Permian*) and sea urchins (*Triassic–today*)
Pelycosaurs (*Late Pennsylvanian–Permian*) and elephants (*Eocene–today*)
Dinosaurs (*Triassic–Cretaceous*) and humans (*Pliocene–today*)

Throughout the over half-billion years of fossil history, animal and plant groups, both numerous and scarce, abruptly appear and disappear at differing times, following different life spans. Additional supporting evidence for evolution is in the past (and present) existence of unique flora and fauna on continents (Australia and South America) and certain island groups (Hawaii, New Zealand, and Madagascar). Geographic isolation is the most credible explanation for the restricted habitats of these confined animals and plants.

Today, anatomical, functional, and superficial changes have been introduced by humans in selective breeding of animals—cattle, dogs, and horses—and in plants—flowers, fruits, and vegetables. In addition, radiometric dating of fossil-bearing sedimentary rocks has confirmed the great expanse of time over which evolution has taken place. Faced with all of this evidence, therefore, most scientists accept evolutionary change as an established fact of nature.

Chapter 7

The Dating Game

Time—
Its Relationship to Past Life and Geological Events

Imagined time travel backwards is a prime passion of astute geologists, particularly paleontologists and historical geologists. Perhaps no other sciences but geology and astronomy encounter the profound role played by the immensity of time. Discriminating the ages of Earth's rocks, many containing entombed significant fossils, is as relevant to geologists as the determination of the age of the solar system and universe is to astronomers.

Because the human life span is far too brief to allow a view of lands sinking, heaving, or moving about, and of oceans changing shape or disappearing, a myopic view of everlasting lands and seas is the popular misconception. The general population knows that the Earth is not quiet, for they experience (or the media informs them of) natural catastrophes such as earthquakes, volcanic eruptions, tsunamis, floods, hurricanes, tornadoes, and wildfires. Nevertheless, inconspicuous snail-paced geologic processes and events go unnoticed—but their cumulative long-term effects are obvious everywhere.

Crustal uplift is obvious where sedimentary rocks with entombed marine fossils crown lofty mountains such as the Alps, Himalayas, and Rockies. Deeply formed plutonic igneous rocks such as granite, gabbro, and anorthosite are exposed on the surface of the Sierra Nevada Mountains of California, the Black Hills of South Dakota, the White Mountains of New Hampshire, and the Adirondacks in New York. Yet in our fleeting lifetime we cannot witness the gradual protracted transformation that brought these original submarine and subcrustal rocks to their present elevated positions.

As an aid to comprehension of the biological and physical events of Earth's history, geologists have subjectively compartmentalized geologic time into:

Eons, the largest time divisions (Azoic, Archean, Proterozoic, Phanerozoic)
Eras, (Aphebian, Helikian, Hadrynian, Paleozoic, Mesozoic, Cenozoic)
Periods, divisions of the eras (ex., Cambrian, Devonian, Jurassic)
Epochs, of periods (ex. Mohawkian, Ulsterian, Pleistocene)
Stages, may be equivalent to or divisions of epochs

These successively finer divisions of geologic time are analogous to the successively finer units of time we use during a calendar year—months, weeks, days, hours, minutes, and seconds. But while every day contains the same number of hours, every geologic time division does not contain the same number of years. For example, the Mesozoic Era (popularly known as "the age of the dinosaurs") encompasses approximately 180 million years. The Paleozoic Era (an era marked by an abundant, but "pre-dinosaur" fossil record) includes almost 300 million years. The reason for this apparently inconvenient system is that geologists find it more useful to create these divisions to encompass *significant physical or biologic events.*

Estimates of the age of the Earth have ranged from 4.6 billion years (based on astronomical, chemical, geological, and physical laws) to approximately 6,000 years (based on Biblical generations). In the past, some rather precise creation dates have been postulated. Adherents to religious fundamentalism give credence to these figures despite tangible scientific evidence to the contrary. In 1598 the noted astronomer Johannes Kepler calculated the time and date of creation to be 11 am on Sunday, April 27, in 3877 bc. No less astonishing is the date arrived at by Anglican Archbishop James Usher, a respected Bible scholar—9 am on Sunday, October 23, in 4004 bc. Today, these purported dates (and times!) for creation seem naïve and perhaps even amusing, but the erudite of the pre-twentieth century could scarcely have done better with the data and instruments available to them.

Discerning the Earth's geologic history can generally be done by applying a number of laws. The *Law of Original Horizontality* says that upon settling on land or in water, transported sediments are deposited as horizontal laminae or "beds," and eventually consolidate into rocks from the weight of overlying sediments in combination with natural cementing agents such as calcite or quartz dissolved in groundwater. This law is relied upon by sedimentologists and stratigraphers (those scientists who specialize in studying sediments and sedimentary rocks) to decipher geological history.

Three additional doctrines are relied upon by geologists to decipher geological history—the *Laws of Superposition, Uniformitarianism, and Paleobiological Succession.* The Law of Superposition states that in undeformed rock strata, newer rocks overlie older ones. This principle is essential for the determination of the relative ages of sedimentary layers. A perhaps more fundamental geological principle is the Law of Uniformitarianism, advanced by Scottish geologist James Hutton in 1785. It states that geological processes (such as erosion and deposition) and natural laws (such as gravitation) now operating to modify the Earth's crust have not changed over the vast stretches of geologic time. This principle is often stated simply as "the present is the key to the past." It is noteworthy that this principle can be verified; when astronomers look deep into space, they also look back in time. Processes such as nuclear fusion and orbital mechanics are observed as they occurred billions of years ago—and appear to have worked exactly as they do today.

Fossil animals and plants succeed themselves in an orderly and determinable sequence. This Law of Paleobiological Succession stipulates that each defined age within the Paleozoic, Mesozoic, and Cenozoic Eras can be identified by their diagnostic fossils. These index fossils are dependable indicators to specified intervals of time. Ideally, they were mobile families, genera or species that were short-lived (relative to geological time), geographically widespread, relatively abundant, and easily identifiable by paleontologists with expertise in specific fossil groups.

The discovery of natural radioactivity near the beginning of the twentieth century permitted the creation of a more precisely dated time scale whereby actual ages, in years, could be assigned to rock units.

The total chronicle of Earth's geologic history cannot be told unless we have a continuum of rocks, preferably fossil-bearing, documenting the events and life of *all* geologic time; it is not surprising that no single locality displays such an all-encompassing continuum. Such a continuum is actually not possible when one considers how geologic processes function. Same-age sediments representing identical environments are not deposited everywhere during a specified segment of time. Whereas an area may undergo relatively rapid uplift and correspondingly rapid erosion at one time—leaving no rock record—millions of years later that same area may be inundated by water, to be followed by formation of new sedimentary rocks. To the geologist, the time gap in the rock record at this location is revealed by an *unconformity* (Fig. 37).

Over limited time scales, it is possible for different strata to be deposited without interruption—the contacts between these strata are termed *conformable.* Since no single locality encompasses rocks from all geologic time, and within a given rock exposure unconformities may exist, the assembly of a complete geologic history can only be accomplished by creating a composite history using data obtained from different regions.

One particularly useful and significant geologic reference section is exposed at the geologically unique Grand Canyon of the Colorado River in northern Arizona. Here, in a single marvelous panorama, over

Fig. 37. Folded erosional surface (unconformity) between tilted Lower Devonian Helderberg Limestone (≈418 mya) on left, and vertical thin-bedded Middle Ordovician Austin Glen graywackes and shales (≈450 mya). Note talus at base. Two mountain-building episodes are demonstrated: the earlier Taconian Orogeny, which deformed the Ordovician strata; and the later Acadian Orogeny, which deformed both Ordovician and Devonian strata. Along northeast side of access road to I-87 from NY 23, northwest of Catskill, Greene County.

one billion years of time is captured in the multi-aged rocks. Even in this awesome spectacle, however, there are major interruptions in the sedimentary sections. We are obligated, therefore, to look elsewhere to locate the rocks that illustrate the events and life of these elusive time gaps. This same procedure must be employed wherever an analysis of geologic history is undertaken.

Eastern New York and environs are no exception. Here, portions of geologic time are unrecorded either because no sediments were laid down or, if they were, the sediments or sedimentary rocks were obliterated by subsequent erosion. During the Taconian, Acadian, and Alleghenyan Orogenies, the processes of tectonic compression, crustal uplift,

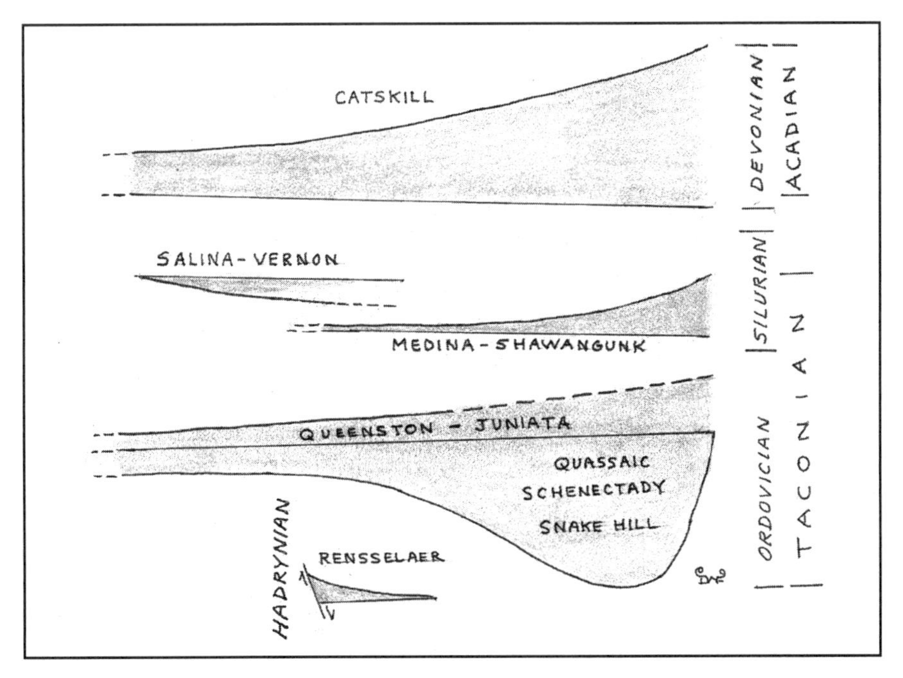

Fig. 38. Clastic wedges—depositional evidence of crustal instability.

and erosion were at their zenith. Within this disturbed arena, deposition was spotty or nonexistent. Instead, erosive debris was carried by water westward, building thick *clastic wedges* (Fig. 38) consisting of expansive deltas and coastal plains west of the present Hudson River (which did not exist during the Paleozoic Era).

The clastic wedges relevant to the topic of this book are listed here in the approximate chronological sequence in which they formed:

NAME	AGE	EST. MAX. THICKNESS
Rensselaer	Late Hadrynian	2,000'
Austin Glen	Medial Ordovician	2,500'
Schenectady	Early–Late Ordovician	5,000'
Quassaic Group	Medial–Late Ordovician	11,700'
Queenston	Late–Late Ordovician	4,000'
Shawangunk-Medina	Early Silurian	2,500'
Salina-Vernon	Late Silurian	1,500'
Tristates Group	Early Devonian	700'
Catskill	Medial and Late Devonian	5,500'

The configurations of these clastic wedges are illustrated on the paleogeographic maps.

Radiometric Dating

It was not until 1896 that a discovery—triggered by an accident in a Paris science laboratory—permitted an assignment of actual ages, in years, to geologic events and structures in chronological order. The central figure in this epochal event was the last of three successive generations of physicists—Antoine Henri Becquerel. He noticed that a photographic glass plate had fogged while near a specimen of uranium-bearing mineral pitchblende. The spoiled emulsion appeared similar to other emulsions that had been exposed to x-rays, but this fogging was much more intense. Endeavoring to determine the source of this unknown energy, Becquerel directed his laboratory assistants, Pierre and Marie Curie, to investigate the properties of pitchblende. Their investigations revealed the presence of two previously unrecognized elements—radium and polonium. Along with uranium, these elements undergo spontaneous radioactive decay; a process by which an unstable "parent" atom transmutes to a stable "daughter" atom by emitting particles and/or energy. In 1903 the trio shared the Nobel Prize in Physics for their joint discovery of natural radioactivity.

Once the actuality of natural radioactivity was established, chemists and physicists clamored to engage in research furthering the development of this novel discovery. British physicist Ernest Rutherford, while teaching at McGill University in Montreal, was a significant contributor. After counting emissions from radium for many days, he noted that total emission activity was proportional to the number of unstable parent atoms still present. This meant that emission must decrease in some regular fashion over time, and that such decay could be expressed mathematically. Also, he suggested that the progressive accumulation of daughter elements—such as lead in uranium-bearing rocks—might provide a means for obtaining a numerical measurement of geologic time. Using Rutherford's suggestion, Yale chemist Bertram Boltwood—who had demonstrated in 1905 that lead was the ultimate disintegration product of uranium—proposed the thesis that a decay series, such as uranium to lead could constitute a means of determining the age of minerals that contained radioactive elements.

Near the onset of the twentieth century, our understanding of natural radioactivity was refined to provide a convincing and relatively straightforward method for measuring geologic time—*radiometric dating*. Physicists learned that the rate of decay of certain elements is fixed and measurable; the proportion of parent and daughter elements can be used to reveal how long they have been present in the rock. Therefore, dates are established by objective measurement and well-determined physical constants. Complications do occur—for example, less massive daughter elements may escape from the internal structure of the rock if it undergoes melting or high temperature metamorphic processes. Therefore, the radiometric age measures the time elapsed since the rock was last heated or melted.

The radiometric timekeepers most widely used by geologists are:

PARENT	→	DAUGHTER
Uranium-238*	→	Lead-206
Uranium-235	→	Lead-207
Thorium-232	→	Lead-208
Rubidium-87	→	Strontium-87
Potassium-40	→	Argon-40
Carbon-14	→	Nitrogen-14

**The number following the element name is the* atomic mass*—the combined number of protons and neutrons in an atom's core, or* nucleus. *The number of protons in a nucleus—the atomic* number*—determines the identity of the element. For example, all uranium atoms contain 92 protons. Varying numbers of neutrons, and therefore varying atomic masses, however, are possible within a given element. These "varieties" are called* isotopes, *and the ratio of parent to daughter elements is often referred to as the* isotopic ratio.

By 1956 scientists had used radiometric dating to estimate an age for the Earth of 3 billion years—a figure that was widely accepted at that time. That same year, Clair Patterson of the University of Chicago showed that radiometric dating of a variety of meteorites resulted in a common age, averaging 4.56 billion years. Assuming that all objects comprising the solar system formed from condensation within a primitive solar nebula at roughly the same time, we conclude that the Earth also has an age of approximately 4.6 billion years. An age of 4.6 billion years for our planet is widely accepted today—but the oldest terrestrial rocks (from Australia) have been radiometrically dated to be "only" about 4.4 billion years old. The apparent discrepancy between these two dates is explained by the fact that radiometric dating tells the length of time that has elapsed *since the rock was last molten*. Geologists and astronomers now believe that the Earth remained molten for a significant portion of its early history, and tectonic forces have repeatedly recycled any original crustal rock into the Earth's interior to form newer rocks.

Without a doubt, radiometric dating has profoundly changed the attitude with which humans view their place in the totality of time.

Fission-Track Dating

This relatively recent dating method is based upon a rare effect of a decay characteristic of some isotopes—namely, the spontaneous splitting (fission) of an atomic nucleus. This phenomenon occurs in uranium-238. For approximately every two million decay emissions of alpha particles (helium nuclei), one atom undergoes fission. As the atomic nucleus splits, the two (and sometimes three) new nuclei repel each other because of their positive charges. During their speedy and energetic separation, they strip electrons from adjacent atoms; the resulting atomic havoc creates ultramicroscopic imperfection lines in minerals. Etching with acid makes the damage created by the fleeing nuclear fragments, denoted *fission-tracks*, visible under an electron microscope. By counting the density of these tracks, an estimate of the number of atoms that have undergone fission can be obtained. Then, using conventional analytic procedures, the amount of unfissioned uranium-238 can be determined. As with other types of radiometric dating, we can calculate the age using the ratio of the original total number of uranium-238 atoms to the number of unfissioned atoms still remaining.

Fission-track dating is almost as precise as other radiometric techniques. Furthermore, the procedure has the advantages of being less expensive to perform, while using ultraminute samples. Its original successes lay in the determination of ages in the range of a few hundred to a few million years. For minerals older than 100 million years, it yields ±10% accuracy.

Geologic Time Scale Condensed within a Calendar Year

To appreciate and better comprehend the immensity of geologic time—4.6 billion years—let's condense it into one calendar year. On this scale, our time intervals become:

1 year	=	**4,600,000,000 years**
1 day	=	**12,600,000 years**
1 hour	=	**525,000 years**
1 minute	=	**8,750 years**
1 second	=	**146 years**

Unabridged		Abridged into a Gregorian Calendar
Azoic Eon	4,600–4,000 mya	**Dec. 31, 12 p.m.** (New Years Eve): birth of the Earth; oldest dated meteorites 4.56 bya
Archean Eon	4,000–2,400 mya	**Feb. 14** (Valentine's Day): oldest dated rocks in North America, ≈ 3.96 bya (billion years ago) *oldest known fossils (cyanobacteria) ≈ 3.46 bya*
Proterozoic Eon	**2,400–543 mya**	
Aphebian Era	2,400–1,600 mya	**July 4** (Independence Day): oldest known glaciation, ≈ 2.25 bya
Helikian Era	1,600–950 mya	**Sept. 27–28:** deformation and metamorphism of Adirondack rocks, 1.23–1.34 bya—*Elzevirian Orogeny* **Oct. 1:** renewed deformation and metamorphism of Adirondack rocks and initial deformation and metamorphism of Hudson and Housatonic Highlands rocks, 1–1.15 bya—*Ottawan Orogeny* **Oct. 6:** granite intrusions into existing Helikian rocks, 945 ± 5 mya (million years ago)
Hadrynian Era	950–543 mya	**Oct. 30–Nov. 1** (Halloween): gathering and fusion of protocontinents into supercontinent Rodinia **Nov. 10–11** (Veteran's Day): Rodinia goes to pieces with the Iapetus Ocean enlarging between Proto-North America and Proto-Eurasia *oldest known multicellular organisms*
Phanerozoic Eon	**543 mya–today**	
Paleozoic Era	543–245 mya	
Cambrian Period	543–489 mya	**Nov. 19–22:** Iapetus Ocean expanding; population explosion with origin of most phyla of invertebrate animals
Ordovician Period	489–442 mya	**Nov. 23:** beginning of closure of Iapetus Ocean—*Penobscot Orogeny* **Nov. 25–27** (Thanksgiving): beginning of the Green and Berkshire mountains during Phase II and formation of the Taconic Mountains during Phases 3–5 of the *Taconian Orogeny* *oldest known fish*

Silurian Period	442–419 mya	**Nov. 28–30:** deposition of Niagara River Gorge strata and gypsum and salt beds in central and western New York *oldest known land plants*
Devonian Period	418–362 mya	**Dec. 1:** deposition of Helderberg Group limestones **Dec. 2–4:** rejuvenation of Taconic Mountains and deposition of Catskill Delta and coastal plain during the three phases of the *Acadian Orogeny* *oldest known amphibians and insects*
Mississippian Period	362–320 mya	**Dec. 5–6:** deposition of Pocono Mountains strata
Pennsylvanian Period	320–285 mya	**Dec. 7–9** (Pearl Harbor Day): collision of Proto-North America with Proto-Africa creating the Allegheny Mountains—*Alleghenyan Orogeny;* granite intrusions and uplifts in New England, extensive jungles and swamps in present West Virginia, Illinois, and Pennsylvania *oldest known reptiles*
Permian Period	285–245 mya	**Dec. 10–12:** welding of all continents into the supercontinent Pangea surrounded by the globe-encircling superocean Panthalassa; salt beds and sandy deserts in present Kansas; continental glaciers in present South Africa, completion of Appalachian Mountain chain *oldest known mammals • extinction of 50% of known life*
Mesozoic Era	**245–65 mya**	
Triassic Period	245–205 mya	**Dec. 12–14:** Pangea begins its dispersal; beginning of the Atlantic Ocean; deposition of Newark Group of terrestrial sediments and outpouring of Palisades basalt; crustal tension produces rifts floored with freshwater lakes from present Nova Scotia to Virginia *oldest known dinosaurs*
Jurassic Period	205–145 mya	**Dec. 15–19:** history obscure throughout present eastern United States and eastern Canada because of lack of rock record; excellent record in western plains of present United States and Canada *proliferation in numbers and kinds of dinosaurs* *oldest known birds*
Cretaceous Period	145–65 mya	**Dec. 20–24:** uplifting Catskill Plateau; flooding of Atlantic Coastal Plain with accompanying deposition of sands, silts, clays, and marls *oldest known flowering plants* **Dec. 25** (Christmas Day): general lowering of world sea level during the Late Cretaceous spells disaster for many coastal shelf inhabitants *ammonoid cephalopods become extinct* **Dec. 26:** Chicxulub Asteroid impacts in the present Gulf of Mexico at about 1 pm, no more living dinosaurs after this hour

Cenozoic Era		**65 mya–now**
Paleogene Period	65–25 mya	continued stream dissection of elevated Catskill Plateau; beginning of Hudson River
Paleocene Epoch	65–58 mya	**Dec. 26, 1 pm–Dec 27, 11 am**
Eocene Epoch	58–37 mya	**Dec. 27, 11 am–Dec. 29, 2 am**
Oligocene Epoch	37–25 mya	**Dec. 29, 2 am–Dec. 30, 1 am**
Neogene Period	25 mya–today	
Miocene Epoch	25–10 mya	**Dec. 30, 1 am–Dec. 31, 4 am:** beginning of doming and uncloaking of Proterozoic rocks of the Adirondack Mountains by erosional stripping of overlying Cambrian and Ordovician sedimentary rocks
Pliocene Epoch	10–2 mya	**Dec. 31, 4 am–Dec. 31, 8 pm:** earliest known humanoids (Australopithicines) ≈3.5 mya; maximum altitude of Adirondack Mountains
Pleistocene Epoch	2 mya–today	**Dec. 31, 8 pm:** earliest known humans (Africa) [hominids] ≈2 mya **Dec. 31, 11:25–11:55 pm:** landscape resculpted and blanketed by continental ice **Dec. 31, 11:55–11:58 pm:** melting of ice sheet and deposition of glacial sediments; permanent ponding of the Great Lakes and Finger Lakes, and formation of Niagara Falls **Dec. 31, 11:58:30 pm:** oldest known human occupation in Hudson Valley **Dec. 31, 11:59 pm:** New York's mammoths and mastodons become extinct; other animals migrate north or south **Dec. 31, 11:59:20 pm:** Mediterranean and Black Sea floods (Noah's Flood)—radiometric date, 5,600 years before present **Dec. 31, 11:59:46 pm:** life of Jesus Christ **Dec. 31, 11:59:51 pm:** Leif Ericson "discovers" North America **Dec. 31, 11:59:59 pm**: American Civil War in progress **Dec. 31, 1/4 second to midnight:** American astronauts walk on our moon (1969) **Dec. 31, 12 pm midnight** (New Year's Eve) TODAY

part two

Geologic History

"Not to know the events which happened before one was born, that is to remain always a boy."
Cicero

For a better understanding of this part, refer to the following in conjunction with the text: Geologic Time Scale (at end of preceding chapter); Fig. 4: Index Map of New York State; Plate 1: Tectonic Map of Columbia County; Paleogeographic Maps of New York State within the following respective chapters

Chapter 8

Precambrian Time Divisions

Introduction

Generally, the further back in time we travel, the less precise is our placement of geologic ages and structural events. Unlike the Phanerozoic Eon chronology (Paleozoic, Mesozoic, and Cenozoic Eras), where radiometric dates may be complemented by fossil correlation, dating of pre-Phanerozoic (i.e., Precambrian) occurrences is solely dependent upon radiometric dating techniques. As refinements in specialized laboratory equipment and sampling techniques improve, and as new chemical and physical geologic "clocks" become available, increasingly accurate dates will enhance our geological chronometer.

To better understand the historical narrative of the Taconic, Catskill, and Adirondack mountains, it is necessary to review the deposition and kinds of rocks that were present prior to the creation of these mountain ranges. The events that transpired before the Phanerozoic Eon set the stage for the constructive and destructive events that followed. On an even larger scale, this geological overview should clarify the evolution of our planetary oasis in space—from its obscure embryonic state to its present condition.

Limits and duration of eons and eras preceding the Phanerozoic Eon are unsettled. Geologists in different continents cling to their own Precambrian nomenclature. Nomenclature differs even within North America. My preference for the Canadian Proterozoic nomenclature rests on rock similarity and extension of New York rocks with those of the Canadian Shield.

EONS / ERAS	Millions of years ago (mya)	Proposed by Stockwell 1964 (used in Canada)	Proposed by James, 1972 (used by U.S. Geological Survey)
PHANEROZOIC EON ↑			
base of Cambrian	543	570 mya	570 mya
Avalonian Orogeny	575		
		Hadrynian	Precambrian Z
LATE PROTEROZOIC ERA			
Ottawan Orogeny	950	880	800 mya
MEDIAL PROTEROZOIC ERA		Neo-Helikian	
Elzevirian Orogeny	1,300	1,280mya	Precambrian Y
		Paleo-Helikian	
Hudsonian–Penokean Orogeny	1,600	1,640mya	1,600mya
EARLY PROTEROZOIC ERA		Aphebian	Precambrian X
Kenoran–Algoman Orogeny	2,400	2,390	2,500mya
ARCHEAN EON		Archean ?	Precambrian W
	4,000		3,700mya
AZOIC EON		↓	(anticipated) Precambrian V
Earth's birthdate	4,600	?	?

A starfield in the constellation Cygnus. The glowing region of hydrogen gas at the center of the photo is commonly referred to as the North American Nebula and is approximately 3,000 light-years from Earth. Photograph by Steve Nightingale.

The Birth of the Earth

Chapter 9

A Nebulous Beginning

The Azoic Eon: The Earliest 600 Million Years

"Equipped with his five senses, man explores the universe around him and calls the adventure Science."

Edwin Powell Hubble

Geologic history begins with the origin of the Earth and the solar system. The solar system is only a small component of a much larger collection of stars, star clusters, dust, and gases—and other planetary systems—collectively termed the Milky Way galaxy. It is estimated to contain as many as 400 billion stars. Our "home galaxy" is approximately 100,000 **light-years** across; our solar system lies about 3/5 the distance from the center. Our galaxy rotates once in approximately 240 million years—"one rotation ago" the first dinosaurs were beginning to evolve. Billions of other galaxies exist in space, with varied shapes and sizes. Spiral, elliptical, and irregular galaxies are a few of the many shapes we see. Our galaxy is believed to have a structure termed a *barred spiral,* with huge sweeping spiral arms thousands of light-years in length, and an elongated central bulge. In a side view, the Milky Way galaxy would have the appearance of two fried eggs "back-to-back."

Distant galaxies, and clusters of galaxies, are found to be receding from each other—much like dots painted on a balloon will move away from each other as the balloon is inflated. By measuring the expansion rate of the universe (this crucial quantity is referred to as the Hubble Constant—named for the astronomer who measured this expansion in the early twentieth century), astronomers can "run the movie backwards" to estimate when the expansion began. Today, most astronomers generally accept that the universe—energy, matter, time, space, and physical laws—came into being approximately 13.7 billion years ago in an event popularly termed "The Big Bang."

For billions of years to follow, atoms of hydrogen and helium collected under mutual gravitational attraction to form galaxies, and within the galaxies, stars formed. Nuclear fusion within stars created elements more massive than helium; when these stars finished their life cycle, many erupted in colossal supernova explosions that "seeded" space with oxygen, carbon, nitrogen, gold, and the remainder of the 92 naturally occurring elements. Some stars ended their lives in a more gentle process whereby microscopic grains of carbon and silicate dust were ejected into space. In either case, later generations of stars (and solar systems such as ours) formed from this recycled "star debris" to begin the cycle again. It is a process that is occurring today throughout our galaxy.

Approximately 4.6 billion years ago, one particular cloud of cold, rarified gas and dust began to contract, rotate, and flatten. Originally one light-year in diameter, 90% of the mass of this *solar nebula* collected toward the center. As this concentrated central portion of the solar nebula collapsed under the influence of gravitational force, its temperature began to climb and this "proto-star" began to emit a dull red glow. The object continued to collapse, eventually reaching a temperature of millions of degrees at the center. Nuclear fusion began in this core, and the object became a new star—our sun. Meanwhile, located at various distances from this newly formed sun, gravitational and electrical forces ("static cling") gathered much of the remaining gas and dust to form *planetesimals*. In turn, these planetesimals collected to form larger *proto-planets.*

These proto-planets were formed from a mixture of hydrogen and helium gas, silicate and metallic grains, and carbon. In addition, molecules of water, ammonia, methane, and other "ices" were present in the proto-planetary mix. Radiation pouring from the sun vaporized the ices (termed *volatiles*) and forced the lighter gases outward from the innermost planet-forming regions. The silicate dust and metallic grains (termed *refractories*) were not vaporized, however, and remained to form the relatively dense and small *terrestrial* planets—Mercury, Venus, Earth, and Mars. Farther away from the sun, ices stayed frozen and collected to form the huge, low-density planets that gravitationally "swept up" vast amounts of hydrogen and helium to become even more massive. These are the *Jovian* planets—Jupiter, Saturn, Uranus, and Neptune. The most distant, small planet Pluto defies our attempts at classification; current thinking is that it may represent an object more closely related to comets than to planets. Between Mars and Jupiter lies a belt of asteroids—small, rocky and metallic bodies that were prevented by Jupiter's gravity field from collecting into a larger planet. The largest asteroid is several hundred miles across; most are much smaller. As more objects are discovered orbiting the sun, astronomers are struggling to develop a more precise definition of the term "planet."

For the first several million years of their existence, the terrestrial planets underwent a period of intense bombardment from smaller remnant proto-planets. The tremendous energy released in these impacts, along with heat generated by radioactive decay and gravitational compression, resulted in an almost complete melting of the Earth. While in this molten state, denser components of the Earth such as iron and nickel sank to deeper levels while less dense components such as silicate materials tended to collect closer to the surface. Thus, our planet's internal structure differentiated into a nickel-iron *core* surrounded by a *mantle* of less dense components. The outermost "skin" of the Earth cooled to form the *crust*; recent radiometric dating of rocks in western Australia suggests that solidification of the crust may have occurred as soon as 200 million years after the formation of the Earth. For a more detailed view of the inferred internal structure of the Earth, see Fig. 15.

Our moon is uniquely large in comparison to the body it orbits. Astronomers theorize that our moon may have formed when a Mars-sized object struck the 50 million-year-old Earth, splashing a good portion of the newly formed mantle into space where it collected to form our one natural satellite. Analysis of the composition of rocks brought back by the Apollo astronauts in the 1960s and 1970s supports this scenario.

The Earth during the Azoic Eon was a hellish place, completely inhospitable to life. The hot surface was under constant bombardment from the leftover debris of the Solar System's formation. These impacts (the scars of which are still preserved on our moon's airless surface) continued to heat the Earth's surface and release massive outpourings of molten rock from the Earth's interior. One rotation of the Earth (what we term one day) is estimated to have taken 10–15 hours, and the moon was approximately 14,000 miles from the Earth (as compared to an average distance of 240,000 miles today). A scarlet-red sun barely filtered through an orange-colored sky containing sulfurous vapors, ash, and dust.

Paleozoic-Proterozoic contact; Potsdam (?) quartz-cobble conglomerate on weathered gneiss (at sledge hammer). Along NY 29 at Kimball's Corners, 8 miles west of Saratoga Springs, Saratoga County.

Building a Basement (3,900–543 million years ago)

Chapter 10

Embryonic Continents and the Origins of Life

The Archean Eon: The Next 1,500 Million Years

The onset of the Archean Eon (sometimes referred to as Archeozoic) was marked by a greatly reduced planetesimal pelting from outer space. The Earth's crust continued to cool. Volcanic outgassing of water vapor, carbon dioxide, nitrogen, and carbon monoxide created an early atmosphere. Astronomers believe that early comet impacts may have supplied additional water to the Earth; comets are largely composed of water ice. As the Earth continued to cool, the atmosphere became increasingly saturated with water vapor and a new phenomenon occurred on Earth—it began to rain. This torrential and long-lasting rain began to fill low-lying, basalt-floored basins, forming the first oceans. Soluble minerals were leached from the rocks and were carried by rivers into these basins; the Archean oceans were salty—as they continue to be today. Scientists are not in agreement on the next steps in the evolution of the Earth's atmosphere. In one scenario, chemical reactions among these volcanic gasses formed molecules of methane and ammonia, after which bolts of lightning continuously flashed through the ammonia- and methane-rich atmosphere—possibly creating amino acids (considered to be the building blocks of life) that accumulated in the oceans.

The oldest known fossils are found in Archean-age rocks. The remains of tiny, threadlike, chlorophyll bearing *cyanobacteria* are found in 3.46 billion-year-old rocks (the Warrawoona Group) located in western Australia. These organisms were to have a profound effect on our planet—through photosynthesis they introduced free oxygen into our atmosphere. More recent Archean metasedimentary rocks (found in northern Canada and Greenland) contain layered structures of trapped sedimentary particles with precipitated calcium carbonate. These stromatolites were formed by matlike colonies of cyanobacteria and, when fossilized, have a cabbagelike appearance. They continue to exist today in highly saline intertidal zones (which are hostile to other forms of life) along Australia's western coast north of Perth.

Archean rocks are almost exclusively metamorphosed from igneous "parent rocks." Rarely, they are derived from clastic sedimentary rocks that show little evidence of transport. While Archean rocks have not been found in the eastern United States, they are extensively exposed within the shield area of North America (Canadian Shield) encircling Hudson Bay, central Labrador, and the Arctic Archipelago. Near Great Slave Lake in northern Canada is the oldest dated rock in North America—the Acasta Gneiss at 3.96 bya. In the United States, exposures of Archean rocks are restricted to the Lake Superior region and the Black Hills of South Dakota and Wyoming.

The *Kenoran-Algoman Orogeny* (a mountain-building event) marked the close of the Archean Eon. In Ontario and Quebec, planed-down extensively deformed metamorphic rocks mark the roots of these ancient mountains.

Chapter 11

A Proterozoic Potpourri

New Rocks, Continental Shields, Mobile Plates, Ice Ages, and Multicellular Life 2,400–543 million years ago

Early Proterozoic—Aphebian Era (2,400-1,600 mya)

At the onset of the Proterozoic Eon, the solar system had been effectively swept clean of most impacting objects. The Earth was "quieting down" and entering a calmer phase, yet occasional impacts continued. North of Toronto, Ontario, (near Sudbury) is a 155-mile-long elliptical scar caused by an impact with an asteroid about 1,850 mya. Today the site is the largest cobalt-nickel deposit in the world.

Beginning in the Aphebian Era, successive pile-ups of tectonic plates embodying large landmasses termed *continental shields* began to appear. Depending on how geologists demarcate them, there are 12–18 shield areas worldwide (Fig. 18). These shields are the metamorphic rock underpinnings upon which the younger Phanerozoic Eon sedimentary rocks were laid down. The Canadian Shield is the foundation platform for the North American tectonic plate; the Adirondack region in New York is a southern extension of this basement. Whether tectonic plates were mobile prior to the Proterozoic Eon is uncertain.

Many new rock types made their appearance during the Aphebian Era. Rounded, well-sorted sedimentary rocks and their metamorphic equivalents—in addition to conglomerates, sandstones, graywackes, mudstones, banded iron formations, dolostones, limestones, and cherts—predominate in these terranes. Ultramafic rocks are rare. South and west of Hudson Bay in Canada lies a bonanza of thick iron oxide deposits. Most, if not all, of our atmosphere's free oxygen was initially "used up" oxidizing these and other similar deposits. Other new rock types include *tillites*—conglomerates formed from the consolidation of glacially transported pebbles, cobbles, and boulders. An example is the Gowganda Conglomerate, found north of Lake Superior, the most ancient evidence for glaciation. It rests on 2,600 million-year-old rocks and is intruded by 2,100 million-year-old igneous rocks.

Medial Proterozoic—Helikian Era (1,600–950 mya)

Continental shields continued to enlarge by crustal plate accretion. Accelerated erosion of surface rocks produced veneers of sediments—ultimately to become sedimentary rocks—on now extensive continental shelves. Volcanic and plutonic igneous activity was forging a new world. Crustal activity and instability increased in tempo, culminating in the reconstitution of earlier igneous and sedimentary rocks into metamorphic rocks. This transformation occurred during the *Elzevirian* and *Ottawan Orogenies* (collectively, these two mountain-building episodes were formerly termed the *Grenville Orogeny*, a name that is still occasionally encountered). Rocks behave plastically while subjected to the tremendous heat and pressure that result in metamorphism; thus, metamorphic rocks frequently exhibit folds that range in size from inches to miles. To further complicate interpretation, original folds are refolded during subsequent metamorphic episodes. Both the Elzevirian and Ottawan Orogenies created contorted rocks exhibiting "ghosts" of older Early Proterozoic structural patterns. A similar metamorphic process was to be repeated over half a billion years later, along present-day New York's border with Connecticut, Massachusetts, and Vermont.

The most extensive Proterozoic landscape region in New York State is the Adirondack area, which can be divided into two distinct provinces, the Adirondack Lowlands and the Adirondack Highlands. Their juncture is a northeast-trending crushed rock zone in St. Lawrence and Lewis counties. This suture, the *Carthage-Colton Mylonite Zone* (Fig. 4), marks the location of a collision or "side-swiping" of two ancient (now inactive) tectonic plates. The lower relief Adirondack Lowlands, occupying Jefferson and St. Lawrence counties, consist primarily of metasedimentary rocks (metaquartzites, calcitic and dolomitic marbles), and metamorphosed impure sandstones containing silicate and carbonate components known as *calc-silicates.* **Charnockites** and leucogranitic gneiss (lacking dark-colored minerals) are found in lesser amounts. Some marbles display the deformed remnants of stromatolite fossils—the oldest known fossils in New York State.

The Adirondack Highlands have a more rugged, forested topography with abundant lakes. The rocks found here consist mainly of meta-plutonic rocks such as **amphibolites,** olivine garnet-bearing metagabbros, and granitic gneisses (Fig. 39). The skyline of southern Franklin and most of Essex counties is dominated by the spectacular McIntyre Range (Fig. 40), sometimes referred to as the "High Peaks" region of the Adirondacks. Mount Marcy is its highest peak at an elevation of 5,344-feet. This range is composed largely of metanorthosite, a rock that is also common on our moon. Lesser quantities of metasedimentary gneisses and marbles are found in the Adirondack Highlands. Non-metamorphosed pegmatites are occasionally encountered in the Highlands. While insignificant in terms of quantity found, these igneous injections are most useful as time markers. Radiometric dating of these bodies indicates that they were emplaced approximately 950 mya, and may represent post-metamorphic "last gasps" of the Ottawan Orogeny. The youngest igneous rocks in the Adirondacks are basalt dikes measuring from a few inches to as large as 15 feet wide. Some are known to be Precambrian in age and are probably associated with a Late Proterozoic thermal event (resulting in the creation of significant igneous bodies) to be described in the next section of this chapter.

Radiometric dates of 1,336 mya to 996 mya for metamorphic rocks found in the Adirondacks demonstrate that colossal transformations (the Elzevirian and Ottawan Orogenies) had been underway for at least 340 million years. Perhaps some of these rocks are even older; in southern Ontario, zircons in metasedimentary rocks are found to be 2,700 million

Fig. 39. Abrupt formational contact between Middle Proterozoic (Helikian) Prospect Mountain granitic gneiss on left, and mafic granulite on right. East side of Wall Street, 4.8 miles north-northwest of I-87 Interchange at Lake George.

Fig. 40. Looking southwest at Lake Placid from the summit of Whiteface Mountain (elevation 4,867 feet); metanorthosite in foreground. Mt. Marcy is the highest peak along the horizon, near the left edge of the photograph.

years old, while some anorthosite found in Labrador is 1,650 million years old. Some Adirondack metamorphic rocks may trace their ancestry back to these ancient Canadian dates.

The Elzevirian and Ottawan Orogenies produced the metamorphic basement upon which later strata would be deposited; this basement underlies most (if not all) of present New York State. In addition to the Adirondack Mountains, the Middle Proterozoic basement is exposed approximately 200 miles south of Mount Marcy. This prominent, NE–SW trending upland region is referred to as the Hudson Highlands and is located in southernmost Dutchess, Putnam, and northernmost Westchester counties. West of the Hudson River, this region extends into southeastern Orange and northwestern Rockland counties. This metamorphic terrane consists of various gneisses (including the Storm King granitic gneiss with a radiometric date of 1,150 mya) along with amphibolites and charnockites. This uprooted massif was thrust-faulted northwestwardly over much younger Middle Ordovician Snake Hill shale and siltstone. Detached erosional remnants of the Hudson Highlands occur sporadically in front of the main thrust fault in Orange County. The Hudson Highlands were displaced during Phase V of the Taconian Orogeny.

Bordering the Hudson Highlands on the south, east of the Hudson River, is yet another metamorphic terrane termed the *Manhattan Prong* (Fig. 4). This intensely folded region consists of the Middle Proterozoic Fordham Gneiss, radiometrically dated to have been last metamorphosed approximately 1,170–1,157 mya (Aleinikoff, 1985). Contained within the Fordham Gneiss are the Upper Proterozoic Pound Ridge and Yonkers granitic gneisses with radiometric ages of 596 ± 19 my and 575 ± 10 my, respectively. These granitic gneisses, in turn, are overlain unconformably by a Lower Cambrian to Middle Ordovician metamorphic sequence of Lowerre Quartzite, Inwood Marble, and Manhattan Schist. These units are equivalent to, respectively, the unmetamorphosed Poughquag Orthoquartzite, Wappinger Dolostone, Balmville Limestone, and the Snake Hill and Walloomsac Shales of the Hudson Valley further to the north. The Hudson Highlands have been altered repeatedly by mountain-building events. While these rocks were unquestionably altered during the Ottawan Orogeny, the degree of overprinting by the subsequent Taconian, Acadian, and Alleghenyan Orogenies has not been convincingly demonstrated. Similarly, the Lower Cambrian through Middle Ordovician rocks of the Manhattan Prong have been folded and metamorphosed—but by which phases of these Paleozoic mountain-building episodes has not been determined.

Other regions of eastern New York and the Mid-Hudson Valley contain lesser exposures of Medial Proterozoic-age rock. In Columbia County, the oldest rock found is an uprooted sliver (the Ghent Block) along the Gallatin Fault, 2.75 miles northeast of Philmont (Plate 1). This NNE–SSW trending exposure is exposed along Arch Bridge Road on the Stottville Quadrangle. The rock is an extremely weathered and crushed plagioclase-rich gneiss overlain by the Poughquag Orthoquartzite and Stissing Dolostone. A similar but larger sliver 5.5 miles to the south comprises Stissing Mountain on the Pine Plains Quadrangle in northern Dutchess County. Here, the sequence is the same—Middle Proterozoic gneiss overlain by

Lower Cambrian Poughquag Orthoquartzite and Stissing Dolostone. Another small sliver of orange, orthoclase-rich granitic gneiss occurs on Todd Hill, along the east side of the Taconic Parkway at its intersection with Todd Hill Road (Pleasant Valley Quadrangle). While the exposures described here are limited, it is assumed that the Middle Proterozoic basement—exposed in the Adirondack Mountains and Hudson Highlands—extensively underlies the more recent rocks covering them in the Mid-Hudson Valley.

Just how deep is this metamorphic rock basement in the Mid-Hudson Valley? To answer this question we will examine the results of an aeromagnetic study (Harwood and Zeitz, 1974). The basement rocks in question contain magnetic minerals such as magnetite that can be detected by a magnetometer carried on an airplane. Successive passes over an area allow the researcher to construct a field map of magnetic anomalies created by this subsurface rock; the intensity of the magnetic field on such plots can be used to infer the depth of the subsurface basement. Analysis of data collected during these overflights revealed two broad, intense subsurface magnetic anomalies beneath the Taconic Mountains—the Albany, New York–Bennington, Vermont anomaly and the Beacon (Newburgh)–Copake anomaly.

The more northerly Albany, New York–Bennington, Vermont anomaly is caused by a subterranean swell of intensely metamorphosed basement, trending at ≈ N 60° E. To the west and north of the Capital District, outcroppings of Late Cambrian through Medial Ordovician continental shelf limestones and dolostones exist. Harwood and Dietz calculated a basement depth of 6,500 feet at a location seven miles southeast of Albany. Northeast of the village of Nassau the depth was determined to be 13,500 feet.

The southerly Beacon (Newburgh)–Copake anomaly is caused by another swelling of intensely metamorphosed magnetic basement, with a trend of ≈N 30° E. Sporadic Middle Proterozoic metamorphic surface slivers of this highly magnetic body are exposed at Stissing Mountain, Corbin Hill, and Todd Hill in Dutchess County and at Cronomer Hill at Newburgh in Orange County. Located at the northwestern margin of the Beacon (Newburgh)–Copake anomaly is a high-angle reverse fault, as displayed on the west side of Stissing Mountain. North and south of this location is a markedly straight high-angle reverse fault—the leading edge of the Ancramdale Slice. Harwood and Dietz calculated the following basement depths for the Beacon (Newburgh)–Copake anomaly at four locations:

- 3,900 feet, 6 miles SSE of Poughkeepsie
- 8,300 feet, 12 miles ENE of Poughkeepsie
- 4,500 feet, 10 miles NE of Poughkeepsie
- 4,100 feet, at a location 3 miles SW of Copake

Late Proterozoic—Hadrynian Era (950–543 mya)

Unlike Early and Medial Proterozoic times, which were characterized almost entirely by the formation of metamorphic rocks, Late Proterozoic time is virtually devoid of them except within restricted sites—for example, the eastern Taconic Mountains. Worldwide, extensive areas display only thick sedimentary rocks, signifying intensive global erosion of earlier rocks. Noticeably subordinate igneous rocks are basalt dikes and sills, and local granite intrusions. Widely spaced glacial deposits dating from 790–650 mya complete the rock suite.

The most significant and unique structural event to occur during the Late Proterozoic was the breakup of the supercontinent *Rodinia*. Rodinia, the largest single continent in history, was dispersed into approximately ten discrete plates by gigantic tensional forces driven by global tectonism. A network of normal faults fractionated these diverging landmasses into rising uplands (*horsts*) and down-dropped lowlands (*grabens*) resulting in landscapes of great relief (Fig. 41). This rugged topography accelerated erosion in the uplands and created correspondingly thick deposition in the lowlands. The clasts contained in these sediments were not transported far—they are angular and poorly sorted (more extensive transport would tend to round and sort them by size). Rodinia's disintegration set in motion the mechanism for the fractured tectonic plates of *Laurentia* (later to become part of North America) and *Baltica* (later to become part of Eurasia) to separate. The enlarging waterway between them was the *Iapetus Ocean*.

Within New York State, the break-up of Rodinia is referred to as the Rensselaerian **Taphrogeny**. The name is derived from the Rensselaer Plateau, the prominent upland region in eastern Rensselaer County, which owes its relatively high elevation to the durabil-

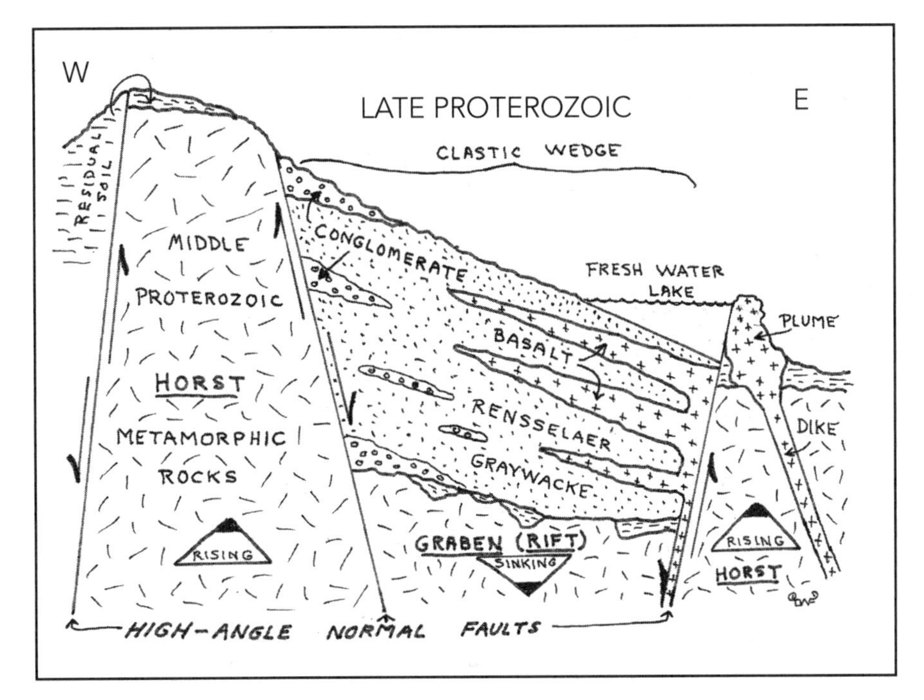

Fig. 41. Late Proterozoic (Hadrynian) horst and graben topography.

ity and resistance of the *Rensselaer Graywacke* bedrock. The piecemeal southward extension of this rock into Columbia County may be observed in the Canaan, Chatham, East Chatham, and State Line quadrangles (Plate 1). The Rensselaer Graywacke (part of an inclusive suite known as the Rensselaer Formation) is presumed to have accumulated as a wedge-shaped talus deposit along the base of a precipice formed by a major high-angle normal fault. This rock unit is at least 1,500 feet thick in Rensselaer County and thins rapidly to the east, where it has been reported to occur as thin, lens-shaped strata within the Everett Schist in western Massachusetts (Zen and Hartshorn, 1966). The direction of sediment thickening, and composition of the clasts (including garnet, zircon, and mica group minerals), indicates that the source area was a metamorphic terrane believed to have been located to the "west." This source region was the Middle Proterozoic basement, today buried under a thick sequence of Lower Paleozoic Cambrian–Ordovician carbonate rocks.

The Rensselaer Formation consists of poorly sorted sedimentary breccias, conglomerates (Fig. 9), graywackes, quartzitic sandstones, and silty, maroon shales. Some of the conglomerates may be glacial in origin; this is a reasonable assumption because Late Proterozoic glaciation has been identified in other countries. All the rock types described above contain veinlets of igneous intrusions containing white quartz (and occasionally, salmon-pink-colored orthoclase feldspar). These "spiderweb" granitic intrusions may be related to larger plutonic intrusions observed in Putnam and Westchester counties. The Poundridge and Yonkers granitic gneisses have radiometric dates of 596 ± 19 mya and 575 ± 10 mya, respectively. If the correlation between these gneisses and the intrusions found in the Rensselaer Graywacke is valid, the Rensselaer Graywacke must be older than the Cambrian, whose base has been universally accepted as 543 million years. Undated ultramafic volcanic rock (termed *greenstone*) flows and dikes occur within the Rensselaer and Austerlitz Formations, to be described in more detail later. Numerous basalt dikes are also known, cutting the Rensselaer and Nassau Formations in Rensselaer and Washington counties (Fig. 37); to date, none of these have been radiometrically dated. In Columbia County, examples of these volcanics are found on Fog Hill and near Mercer Mountain (State Line Quadrangle) and north of Queechy Lake (Canaan Quadrangle) (Plate 1). Similarly, numerous undated unmetamorphosed basalt dikes penetrate 1,100 million-year-old metamorphic rocks in the Adirondack Mountains; geologists favor an age of approximately 600 million years for these intrusions. One example is the six-foot-wide dike at the "Noses" between Fonda and Canajoharie in the Mohawk Valley that cuts across Middle Proterozoic garnet-rich gneiss. Potsdam Sandstone and Little Falls Dolostone, both Late Cambrian age, overlie the dike.

If the Taconic dikes are the same age, this allows for an interval of approximately 100 my for basalt dike injection. Until the Taconic dikes and greenstone flows are radiometrically dated, we cannot be certain whether they were injected as a single event or several events during the Later Proterozoic, Early Cambrian, or Medial Cambrian ages.

Although occurring in separate thrust slices, the Austerlitz (green and purple phyllites, maroon sandstones, and quartzites), Elizaville (dark gray-green argillite and interbedded quartzites), and Everett (schist, quartzite, and gray-green phyllites) Formations are probably linked sedimentologically. All may be metamorphosed equivalents of rocks found in the Nassau Formation. The Nassau Formation is found within the Kinderhook, Ravena, Hudson North, Hudson South, Stottville, and East Chatham Quadrangles in the "Low Taconics." Within the "High Taconics," the Elizaville and Austerlitz Formations are in a more easterly belt centered on the Taconic State Parkway. The Everett Formation is confined to the hills flanking the New York–Massachusetts border (Plate 1).

To date, no indisputable key fossils have been reported from the above-mentioned formations except for a bonafide Early Cambrian trilobite fauna in the uppermost 200 feet of the Nassau Formation. Exceedingly rare, vague trace fossils (*Oldhamia*) have been seen in the maroon-purple shales of the Rensselaer and Austerlitz Formations. Five-rayed, gray-black impressions of an unidentified animal (*Ediacaran* [pr. ee-dee-ak'-a-ran] fauna?) have been found in the Elizaville Formation. Lacking conclusive evidence for a Cambrian age, I choose to assign the Rensselaer, Austerlitz, Elizaville, Everett, and most of the Nassau Formation to the Late Proterozoic (Hadrynian) age.

Usually, there is a pronounced angular unconformity between the basal Paleozoic (commonly Cambrian or Ordovician) strata and the underlying, highly deformed, metamorphosed Middle Proterozoic rocks (Fig. 43), as in the Adirondack Mountains. Where Upper Proterozoic sedimentary rocks intervene, however, structural discontinuity at the base of the Cambrian rock is rare. As a result, the demarcation between Late Proterozoic and Paleozoic time depends upon the placement of the earliest appearance of phosphatic shelly fossils and, to some workers, the earliest trilobites. Within a thick sequence of conformable strata spanning this boundary, in lieu of diagnostic fossil evidence, the placement of the Proterozoic–Cambrian contact becomes ambiguous.

In the Avalon Peninsula of Newfoundland, fossiliferous Lower Cambrian strata truncate a granite (dated at 575 mya) that, in turn, penetrates a thick series of Upper Proterozoic

Fig. 42. Upper Proterozoic basalt dike cutting Middle Proterozoic Bryant Lake Marble (with inclusions) with Hague Gneiss on extreme left. West side of US 4, 1.5 miles north of intersection of US 4 and NY 22 at Comstock, Washington County.

Fig. 43. Paleozoic-Proterozoic contact; Potsdam (?) quartz-cobble conglomerate on weathered gneiss (at sledge hammer). Along NY 29 at Kimball's Corners, 8 miles west of Saratoga Springs, Saratoga County.

sedimentary rocks. Granites of similar age are known throughout the eastern Appalachian ranges southward to Georgia. The Poundridge and Yonkers granitic gneisses fall into this category. Rodgers (1967) and Lilly (1969) associated this plutonic activity with the initial stage of the *Avalonian Orogeny.* Avalon, a micro-tectonic plate, was part of an archipelago in the Iapetus Ocean within which there lived many of the early known forms of invertebrate life. Today, remnants of Avalon have been identified in Nova Scotia, New Brunswick, the Boston Basin, eastern Rhode Island, and North Carolina. These remnants may also be hidden within the Upper Proterozoic strata of the Taconics. The accretion of this archipelago onto the North American Plate will be discussed in more detail in a later chapter (The Taconic Resurrection).

Besides the previously mentioned stromatolites, the Late Proterozoic is notable for the appearance of multi-cellular animals (metazoans). Evolving from simpler one-celled animals (protozoans), these soft-bodied forms are discoidal, frond-like, or radial in shape. They were discovered in 1946 when Australian geologist R. C. Sprigg was prospecting for ore in the Flinders Range, north of Adelaide, South Australia. While enjoying lunch, Sprigg playfully flipped over slabs of Upper Proterozoic Rawnsley Quartzite around him. An unusual natural pattern caught his eye and led to the uncovering of several strange-looking fossils. He thought them to be the remains of jellyfish, but paleontology was outside his scope of expertise. He took them to Martin Glaessner, a noted Australian paleontologist, and the discovery was pronounced to be a new Precambrian fauna (termed Ediacaran) of soft-bodied creatures. The Ediacaran fauna appeared approximately 600 mya and persisted until at least 560 mya. This is Earth's oldest record of the evolutionary radiation of multicellular animals. Because the Ediacaran fauna is extremely rare and the assignment of its enigmatic members is taxonomically controversial, they are not considered to be reliable index fossils. Nevertheless, their discovery sites indicate a very late Proterozoic age.

Trilobite *Basidechenella rowi* from the Medial Devonian Moscow Formation; collected north of Letchworth State Park at York, Livingston County. The existence of trilobites spanned the entire Paleozoic Era. Size of each approximately 1 inch (2.5 cm). Photo by Steve Nightingale.

Three Hundred Million Years of Growing: The Paleozoic Era (543–245 million years ago)

Chapter 12

The Cambrian Menagerie—A Population "Explosion"

543–489 million years ago

During Cambrian[4] time, continental configuration and distribution were very different from that of today. Proto-North America (*Laurentia*) and earliest Scandinavia (*Baltica*) were east of Proto-South America. All three of these landmasses were located in the southern hemisphere, between 0° and 30° south latitude, and encircled the expanding Iapetus Sea, which had its origin during the Late Proterozoic Era.

For the first 50–55 million years of the Paleozoic Era, the continents displayed a lifeless, Mars-like wasteland. The marine environment, by contrast, was host to a widespread explosion of invertebrate life. Virtually *all phyla* (but not *classes*) of invertebrate animals made their first appearance during the Cambrian Period. The majority of these creatures were alien to any we find today and left a sparse fossil record that is frustratingly difficult to interpret. Many Cambrian creatures were so anatomically unfamiliar that paleontologist-taxonomists are baffled as to their proper classification relative to modern marine animals. As a result, these mysterious, bizarre, and mystically functioning sea-dwellers have been "pigeonholed" into classifications that are not represented in subsequent geologic ages. These creatures are best exemplified by the *Burgess Fauna*, discovered by Charles D. Walcott in the Middle Cambrian Burgess Shale in British Columbia. The books *Wonderful Life* (Gould, 1989) and *The Fossils of the Burgess Shale* (Briggs et al., 1994) present delightfully entertaining overviews of this enigmatic population.

Among collectors of fossils, the most popular and easily identifiable featured players of the Cambrian cast are the trilobites. This extinct class of bottom-dwelling crustacean–arthropods functioned as scavengers, consuming decaying plant and animal litter on the sea floor. They were so numerous and diversified that the marine continental margins were dominated by them. Trilobites make excellent index fossils, and provide the stratigraphic zonation by which the Cambrian Period can be resolved into finer time divisions. From the limestones and shales of the Germantown Formation—exposed in various locations in Columbia County—the following seven trilobite zones have been identified (illustrations and descriptions can be found in the *Treatise on Invertebrate Paleontology* Vol O (Arthropoda I), 1959, Raymond C. Moore, editor):

Late Cambrian	— *Ideomesus*
	— *Aphelaspis*
Medial Cambrian	— *Hypagnostus, Centropleura*
	— *Ptychagnostus gibbus*
Early Cambrian	— *Pagetides*
	— *Acimetopus*
	— *Elliptocephala, Microdiscus* (both also in Nassau Formation)

Trilobites periodically shed (molted) their chitinous or calcium phosphate (phosphatic) exoskeletons as their soft bodies grew—as crabs and lobsters do today. Sometimes bedding planes are strewn only with fossilized head (*cephalon*), tail (*pygidium*), or body (*thorax*) segments.

Within Cambrian intertidal zones (Fig. 44) stromatolites were conspicuous. They constructed barrier

Fig. 44. Ripple marks in Potsdam Sandstone near Lake George shore, Hearthstone Point Campsite, 2 miles northeast of I-87 Interchange at Lake George.

Fig. 45. Stromatolites, glacially planed to show cross sections resembling cabbages or Brussels sprouts. This fossilized reef is in the Upper Cambrian Hoyt Limestone at the "Petrified Gardens" on Petrified Gardens Road, 3 miles west of Saratoga Springs, Saratoga County.

blanket reefs whose interstices provided a tranquil haven for other intertidal dwellers against surging surf and tides. Two of the finest exposures of fossil stromatolites may be seen three miles west of Saratoga Springs on Petrified Gardens Road, North of NY 29 (Fig. 45). An extensive exposure is found at the privately owned "Petrified Sea Gardens," where three separate blanket reefs in the Upper Cambrian Hoyt Limestone may be viewed and photographed. A second, less extensive, locality is at nearby Lester Park, owned by the New York State Museum (see Goldring [1938] and Fisher [1991]). Collecting is strictly forbidden at both sites.

In Columbia and Dutchess counties fossil evidence indicates that the uppermost 200 feet of the Nassau Formation and the lower 80–90% of the Germantown Formation are Cambrian in age. The olive-green to gray shales and **argillites**, sandstones, and limestone conglomerates of the uppermost Nassau Formation are Early Cambrian (Fig. 46). The base of the Germantown Formation is a hard, massive orthoquartzite known as the *Diamond Rock Quartzite*. It is well displayed on the east side of the "Castleton Cutoff" railroad cut 1.9 miles south of Schodack Landing. Another exposure is on the north side of Columbia County 14, south of the Olana State Historic Site. At both outcroppings, the contact with the Nassau Formation is exposed.

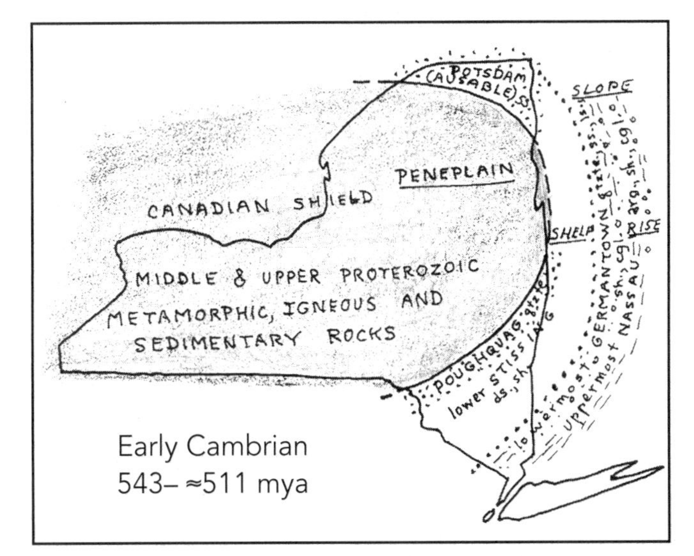

Fig. 46. Paleogeographic map of New York during the Early Cambrian.

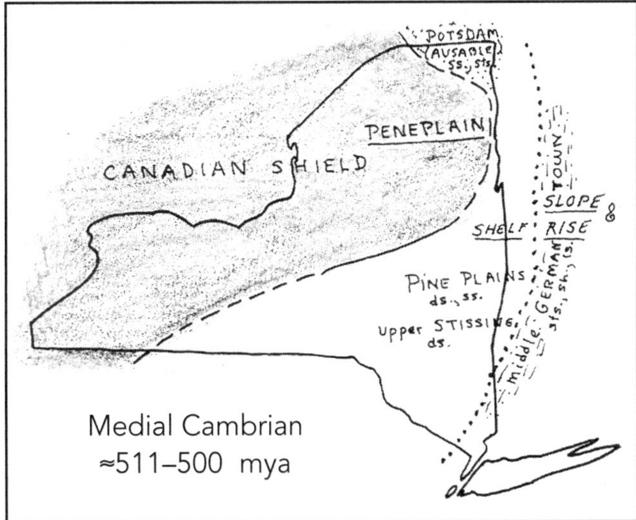

Fig. 47. Paleogeographic map of New York during the Medial Cambrian.

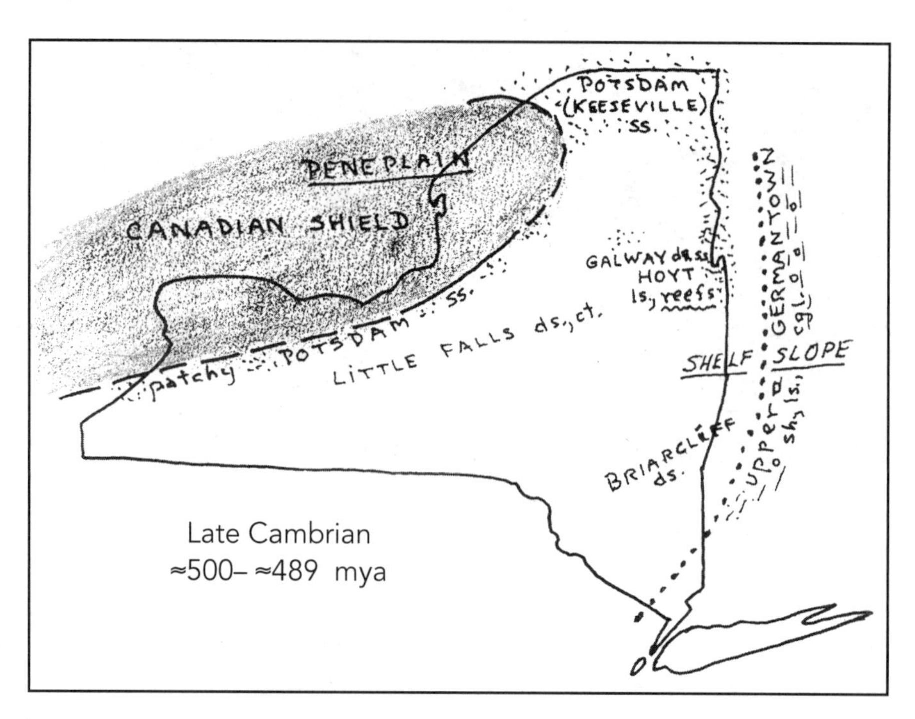

Fig. 48. Paleogeographic map of New York during the Late Cambrian.

Black to gray silty shales, ribbon limestones, and limestone conglomerates of the Germantown Formation have Early, Medial (Fig. 47), and Late Cambrian ages (Fig. 48). Within the Early Cambrian, small and sightless eodiscid (*Pagetides*) and ollenellid (*Elliptocephala asaphoides*) trilobites are rarely found. In addition, the problematica[5] *Hyolithes* and *Hyolithellus, Coleolus,* and phosphatic-shelled brachiopods (lingulids) exist. These may be recovered from the Stuyvesant Conglomerate Member of the Nassau Formation by acetic or formic acid disintegration of the matrix.

Contemporaneous with the Nassau and Germantown continental slope and continental rise deposits were those of the more "westerly" shallow water continental shelf (Fig. 50)—the lower units of the Wappinger Group. These are supra-tidal, intertidal, and subtidal mixtures of quartz sand and calcium-magnesium carbonates.

The "ground floor" of New York's "skyscraper" of Paleozoic rocks—to be built upon the earlier basement of Middle Proterozoic metamorphic rocks—is the Poughquag Orthoquartzite, originally an Early Cambrian beach sand. The second, third, and fourth floors comprise the Stissing, Pine Plains (Fig. 51), and Briarcliff Formations, all included within the Lower Wappinger Group and exposed in Columbia and Dutchess counties. These are sandy shales, dolostones, and quartzose and calcitic dolostones; fossils are exceedingly rare in these units. Other geologists and I have found Early Cambrian trilobites in the Poughquag Orthoquartzite in Dutchess County; Knopf (1962) reported Medial Cambrian fossils from the Stissing Formation and Late Cambrian trilobites from the Briarcliff Formation near Pine Plains. Except for stromatolites, no diagnostic fossils have been found in the Pine Plains Formation.

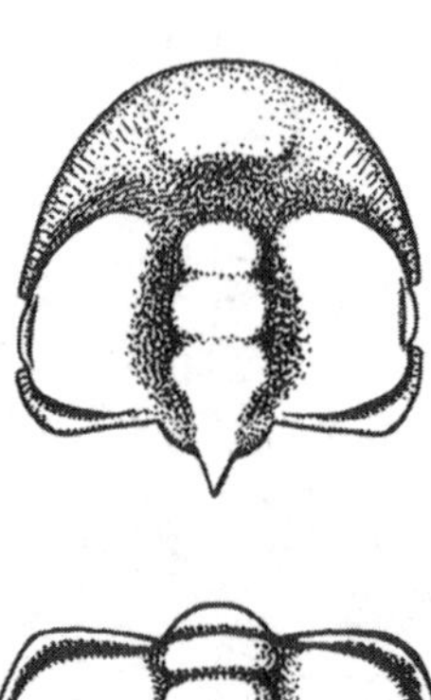

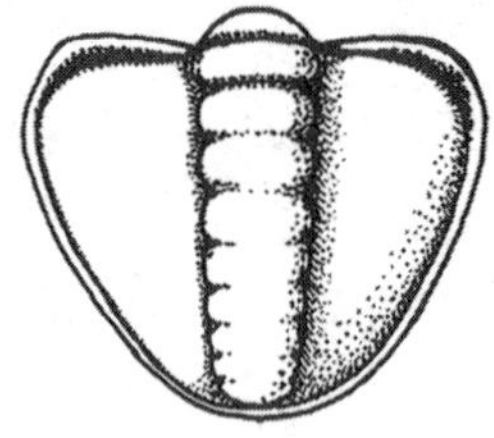

Fig. 49. Early Cambrian eodiscid trilobite, Nassau and Germantown Formations.

Why so few fossils in the Wappinger Group? Their scarcity may be attributed to destruction during the processs of dolomitization—the

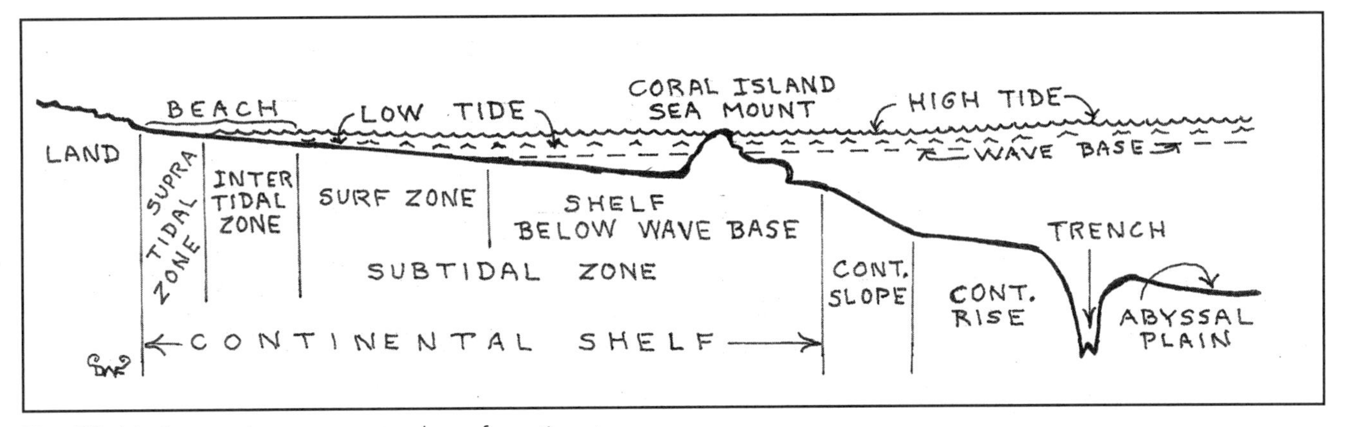

Fig. 50. Marine environments at edge of continent.

alteration of limestone to dolostone. Another reason for a fossil famine may be the widespread existence of seas with abnormally high salinities that would be hostile to most marine life. The presence of primary dolostones, which form under these conditions, argues for this probability. Also, half a billion years ago our moon was closer to the Earth. This would have caused more extreme tidal ranges, ebbing and flowing with great surface agitation upon much broader continental coastal margins.

The Wappinger Group, Nassau, and Germantown Formations all have been transported to various degrees. Rock units that have been transported out of their depositional realm are termed **allochthons**. Rock units that have been transported to a lesser degree, remaining within their original depositional realm are termed **parautochthons**. The allochthonous Nassau and Germantown Formations occur only within the gravity slides and earliest thrust slices (*Livingston* and *Stottville*), which were later high-angle reverse-faulted within the "Low Taconics" (Plate 1). The age-equivalent parautochthonous Lower Wappinger Group (Poughquag, Stissing, Pine Plains, and Briarcliff Formations) appears mainly in the north-south valley occupied by NY 22 and as fault slivers along the leading edges of rock-over-rock slices (Plate 1). While the Wappinger Group continental shelf strata are found today *east* of their contemporaneous continental rise and continental slope strata of the Nassau and Germantown Formations, they were originally located to the *west* of them. The present reverse juxtaposition came about because the continental slope and rise rocks were thrust-faulted "westward" out of their depositional realm and even further "west" over the shelf-rocks synclinorium sequence. Although similarly reverse-faulted, the Wappinger Group shelf rocks remained within their depositional realm (see Plate 1 and the following chapter on the "Restless Ordovician").

Fig. 51. West limb of syncline in Pine Plains Dolostone, north side of I-84, southern Dutchess County.

Chapter 13

The Restless Ordovician, and the Penobscot and Taconian Orogenies: A Mid-Life Crisis

489–442 million years ago

Overview

A sedate scene of crustal stability marked the first several million years of the Ordovician[6] Period in "easternmost New York" and "westernmost New England." Holdover Late Cambrian environments and sedimentation types continued virtually unchanged into early Ordovician time. Marine animals continued evolving, however, bringing forth new taxa of trilobites, gastropods, nautiloid cephalopods, brachiopods, and tooth-like **conodonts**. While this shelly fauna thrived in the shallow water environments, floating **graptolites** (Fig. 14) became entrapped in the deeper oceanic muddy habitats. As time passed, conodonts and graptolites diversified to the extent that they, as index fossils, became useful as the biostratigraphic framework of the Ordovician Period.

While habitats became physically more varied and enticing for food and refuge, newly evolved sea creatures became entrenched as rivals or helpers in the new ecosystems. Among these newcomers were sponges, bryozoans, pelycypods (clams, oysters, scallops), **stromatoporoids**, starfishes, corals, **cystoids**, crinoids, and **ostracodes**. Not to be outdone, the ubiquitous trilobites continued their heyday from the Late Cambrian through the Early Ordovician, occupying most habitable marine domains. Trilobites would eventually decline during Medial–Late Ordovician time in proportion to the increase in nautiloid cephalopods; in later Devonian time, trilobite numbers continued to dwindle in proportion to increased numbers of fishes and cephalopods. Apparently, hordes of cephalopods and fish predators effortlessly preyed on the ill-fated trilobites as tasty morsels.

While the Ordovician biological world was enjoying a relatively unmolested paradise, the physical world was punctuated by structural disarrangement. Beginning near the close of the Early Ordovician, this crustal reconstruction (the *Penobscot Orogeny*) was brought about by a reversal of movement of the Laurentia and Baltica tectonic plates—from *divergence* to *convergence*—and the ultimate collision later in the Paleozoic Era. Here begins the complicated history of events associated with the Taconic Mountains.

The most violent and uplifting periods in the evolution of the Taconic Mountains happened during the Taconian and Acadian mountain-building episodes. A synopsis of the earlier orogenies will be discussed here, while a more detailed account of the later Acadian Orogeny will appear in the chapter "The Taconic Resurrection and the Devonian Renaissance."

ORDOVICIAN GRAPTOLITE ZONES IN NEW YORK STATE

sequence by Professor John Riva, Laval University, Ste.-Foy, Quebec

AGE	GENUS, SPECIES*	SHALE FORMATION	OROGENY	
442 million years ago			PHASE	
	Dicellograptus complanatus	o Queenston o Oswego	V	T
	Amplexograptus manitoulensis	Pulaski Whetstone Gulf		A
	Climacograptus pygmaeus	Frankfort	IV	
	C. spiniferus	upper Utica Schenectady		C
	Orthograptus ruedemanni	lower Utica Dolgeville upper Snake Hill	III	O
	Corynoides americanus	Austin Glen (Syn. Seq.) lower Snake Hill Poughkeepsie		N I
	Diplograptus multidens	Austin Glen (Tac. Seq.)	II	
	Nemagraptus gracilis	Mt. Merino Indian River	I	A
5–10 million year hiatus	x x x x x x x x x x x x x x x x x x x x x x			N
/\			**PENOBSCOT**	
	Glyptograptus dentatus *Tetragraptus fruticosus* *T. approximatus*	upper Stuyvesant Falls (not reported) lower Stuyvesant Falls		
	Rhabdinopora flabelliforma	uppermost Germantown, Schaghticoke		
489 million years ago				

o No graptolites in formation

* For fossil identification, see *Treatise on Invertebrate Paleontology*, Vol V, Graptolithina, 101 pp. (1955)

Chapter 14

Early Ordovician and the Penobscot[7] Orogeny

489–480 million years ago

Since the breakup of Rodinia during the Late Proterozoic Era, the scattered tectonic plates continued their dispersal during the Cambrian and Early Ordovician Periods. Crustal participants in the Taconic theater were Laurentia (later to become part of North America) and Baltica (later to become part of Eurasia). Slower plate separation initially invoked structural quiescence causing similar (and in some cases, *identical*) sedimentary rock types to straddle the Cambrian–Ordovician time boundary.

Sedimentary similarity, however, did not deter organic evolution in marine animals such as trilobites, gastropods, and conodonts. A case in point, the Germantown Formation consists of interbedded silty, platy, dark gray to black shales, laminated ribbon limestones, and sandy dolostone-limestone conglomerates throughout its thickness. Within the uppermost 70–140 feet, the trilobite *Clelandia* and the graptolite *Rhabdinopora flabelliforma* (formerly *Dictyonema flabelliforme*) have been found, indicating an Early Ordovician age. The lower Germantown Formation (at least 600 feet thick) yields Late, Medial, and Early Cambrian trilobites. The Germantown Formation accumulated on the far edge of the continental shelf and continental slope harboring submarine canyons, and possibly extending to the continental rise (Fig. 50). Resting on the Germantown Formation is the Stuyvesant Falls Formation—pale green and greenish-gray shales, green-black cherts, and thin-bedded laminated siltstones and orthoquartzites (Fig. 53). In the past this unit was often confused with the older green shales and sandstones of the Nassau Formation. Interlayered chert beds, abundant trace fossils (supposed worm and trilobite trails), and rare Early Ordovician graptolites, coupled with this unit's position directly below the younger Indian River Formation, however, establish positive recognition of the Stuyvesant Falls Formation. The environmental realm of the Stuyvesant Falls Formation is uncertain. Lack of a bottom-dwelling molluscan-brachiopod fauna and the presence of radiolarian-bearing chert argues for a deeper-water origin.

Likewise, the upper Wappinger carbonate rocks (Halcyon Lake, Rochdale, and Copake Formations)

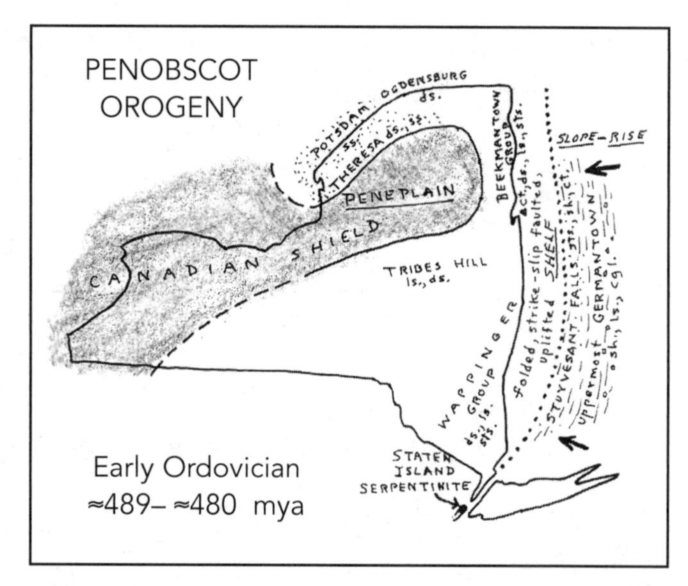

Fig. 52. Paleogeographic map of New York during the Early Ordovician.

Fig. 53. Thin-bedded siltstones and shales, Stuyvesant Falls Formation, Taconic State Parkway, Dutchess County.

are verified as Early Ordovician on the basis of newer trilobites, gastropods (Fig. 54), nautiloid cephalopods, and conodonts, whereas the lower Late Cambrian Wappinger Group (Poughquag, Stissing, Pine Plains, and Briarcliff Formations) contain none of the Early Ordovician species. Upper Wappinger dolostones and limestones represent nearshore continental shelf deposits. Widespread shallow Late Cambrian and Early Ordovician intermixed carbonate rocks throughout much of the United States attest to the expansive, flooded, very low relief coastal plains during this interval. Such extensive shoals were probably a result of more powerful tides, caused by the lesser distance to the moon.

Although there is no evidence of a structural break between the Cambrian and Ordovician continental shelf deposits in "easternmost New York" and "westernmost New England," the Wappinger Group is capped by a major erosional surface. This *Penobscot Unconformity* was generated by broad folding, minor strike-slip faulting, and regional uplift of the *Penobscot Orogeny* (Fig. 52). As a result, the overlying horizontal Middle Ordovician Balmville Limestone truncates differing Wappinger units. A post-Wappinger, pre-Balmville residual soil consisted of patchy red, orange, and brown iron oxides, which produced the **goethite, limonite, hematite,** and **siderite** iron ores to be discussed in the section, "Economic Resources from the Ground in Columbia County."

The north-central portion of Staten Island exposes a segment of oceanic crust, the *Staten Island Serpentinite*. This marks the initial evidence of Penobscot Orogeny plate convergence.

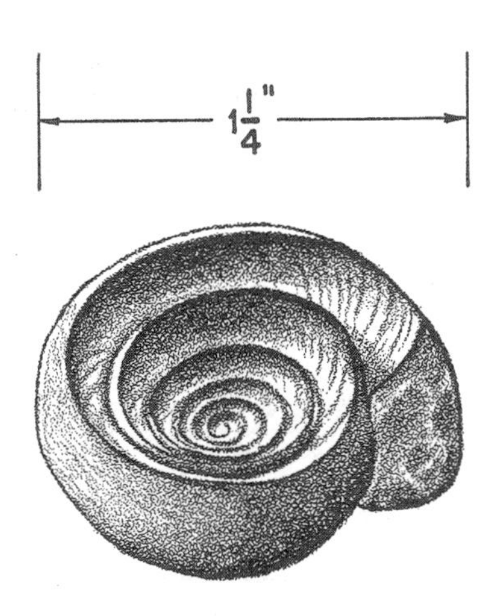

Fig. 54. A depressed spire gastropod (*Lecanospira compacta*), index fossil of middle Early Ordovician age, Rochdale Limestone of Wappinger Group. Printed with permission of the New York State Museum, Albany, NY.

Chapter 15

Early–Medial Ordovician and the Taconian Orogeny (Phase I)

470–465 million years ago

When the ocean resumed flooding upon the Penobscot Unconformity, the now-stable continental shelf spawned a new cast of marine invertebrates in these new and different environments. Evidence of this *Chazy Tranquillity* is best exhibited in the northern Champlain Valley, where several different limestones of the Chazy Group—the Day Point, Crown Point, and Valcour Formations—are found. Together, these formations span a thickness of approximately 700 feet. The Day Point Formation displays prominent **crossbedding** (Fig. 56), which is uncharacteristic for limestone; the crossbeds show a dip of 17–24° east, which implies a nearby western shoreline. In addition to the crossbeds, fossil brachiopods in the formation display size sorting—evidence for highly turbulent surf action. Dome-shaped reefs consisting of algae, bryozoans, sponges, and stromatoporoids abound in the upper Day Point and lower Valcour Limestones (Fig. 57). Because reefs can only form on stable sea floors, the presence of these organisms provides evidence for crustal stability. Fossil fragmental off-reef and inter-reef limestones and fine- to medium-textured, evenly bedded limestones yield a profuse and diverse shelf fauna including nautiloid cephalopods, gastropods, brachiopods, ostracodes, and forty-eight species of trilobites (Shaw, 1968). Deserving special mention is a relatively large (up to 4.5 inches in diameter) flat-coiled snail (*Maclurites magnus*) (Fig. 58); on some bedding planes in the Crown Point Limestone, it is

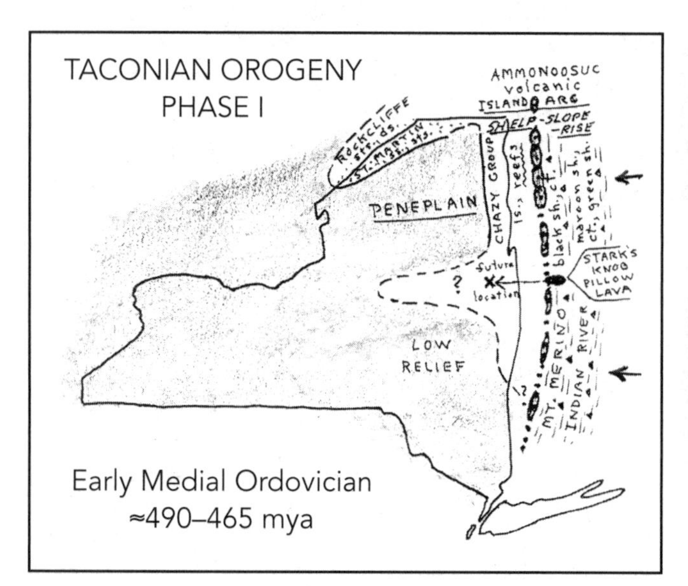

Fig. 55. Taconian Orogeny (Phase I).

Fig. 56. Cross-bedded Day Point Limestone, Souza Farm, south of Chazy village and east of US 9, Clinton County.

Fig. 57. White mounds of algal-bryozoan-sponge-stromatoporoid reefs in horizontally bedded Valcour Limestone, inactive quarry north of Sheldon Lane, 2 miles southeast of Chazy, Clinton County.

Fig. 58. Gastropod (*Maclurites magnus*). First New York State fossil to receive Linnaean nomenclature. Middle Ordovician Crown Point Limestone, near Chazy, Clinton County. Printed with permission of the New York State Museum, Albany, NY.

present to the almost complete exclusion of other species. This gastropod was the first New York fossil to be assigned Linnaean binomial nomenclature; in 1817 the French conchologist Charles A. LeSeur performed this taxonomic assignment. For additional details on Chazyan limestones see Oxley and Kay (1959), Fisher (1968), and Shaw's monograph (1968) on Chazy trilobites.

The onset of crustal unrest becomes evident with the introduction of clastics (clay and sand) in the upper third of the Valcour Limestone, particularly the uppermost 45 feet, which is a calcareous mudstone with clayey and siliceous dolostones indistinguishable from the basal Black River (Pamelia Formation) lithologies on the western flank of the Adirondacks. Phase I of the Taconian Orogeny was now in progress. Meanwhile, off the shelf in deeper waters, Indian River cherts and shales (maroon and green) and Mt. Merino black shales and cherts were piling up. It is significant that both the basal Indian River and the basal Chazy Group (Day Point Formation) exhibit maroon coloration. This coincidental occurrence suggests incorporation of hematite, from the Early Ordovician soil found atop the Beekmantown and Wappinger carbonate groups. Evidence of crustal turmoil is also apparent in the existence of contemporaneous Ammonusuc volcanic flows in Vermont and Massachusetts, and the unique pillow lava (Stark's Knob) located in Northumberland, north of Schuylerville in Saratoga County (Fig. 59). These are examples of new oceanic crust emanating from a submarine active volcanic ridge that later emerged as a volcanic island arc.

Today, no Chazyan formations exist in the Black River, Mohawk, or Hudson Valleys. During the Early–Medial Ordovician, however, Chazyan limestones had to have been present in the Taconic area because unquestionable Chazyan fossils have been found in pebbles and cobbles within the Poughkeepsie Mélange, which accumulated during Phase III of the Taconian Orogeny (Fig. 15). If Chazyan strata were laid down in the Black River and Mohawk Valleys, erosional stripping of them had to occur prior to the deposition of the Black River Group carbonates that rest directly on Middle Proterozoic metamorphic rocks in the Black River Valley and on Lower Ordovician Tribes Hill dolostones in the Mohawk Valley.

Fig. 59. Pillow lava, Stark's Knob, Northumberland, Saratoga County.

Chapter 16

Medial Ordovician and the Taconian Orogeny (Phase II)

465–462 million years ago

Three formations in the Black River Group—the Pamelia Dolostone, Lowville Limestone, and Watertown Limestone—display their greatest development in Lewis and Jefferson counties, where the thickness ranges from 150–275 feet. These limestones, which formed during the *Black River Tranquility,* are represented in the Mohawk Valley by thin, discontinuous strata of two formations in the Black River Group—the Lowville and Amsterdam limestones—that range in thickness from 0–20 feet. In the Champlain Valley the Lowville, Isle LaMotte (Fig. 61), and Orwell Limestones represent the Black River Group; the thickness there ranges from 0–70 feet. In the Hudson Valley, Black River Group formations are absent between the Wappinger Group below and the Balmville Limestone-Conglomerate above. The fact that Black River fossils occur in pebbles and cobbles of the Poughkeepsie Mélange confirms a pre-erosional presence of Black River strata before the formation of the mélange. A limestone exposure, about 12 feet thick along the former railroad in Pleasant Valley, Dutchess County, may be a Black River **relict.**

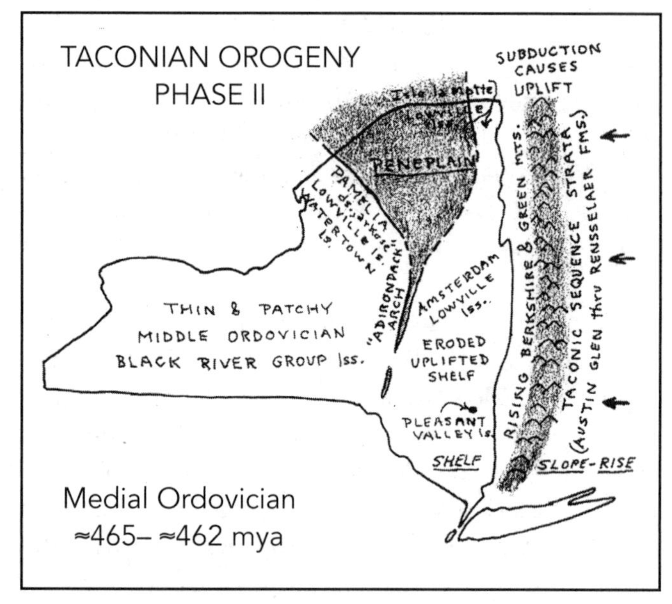

Fig. 60. Taconian Orogeny (Phase II).

This interval of partial to complete erosion of Black River Group Limestones is associated with the development of a *foreland swell,* a term use to describe the uplift, folding, and shortening of the continental shelf rocks. Tectonic plate convergence and subduction create foreland swells; the structure referred to here is known as the *Adirondack Arch*. The Adirondack Arch was a north-south island or peninsula of Tribes Hill dolostones, later covered by Trenton limestones or Utica Shale. Small hills made their appearance in a newly created linear igneous-metamorphic terrane—upthrust by bordering reverse faults that are typically associated with crustal compression. These hills would eventually become the Green and Berkshire mountains. The *Taconic Sequence* abutted the Berkshires on the east.

Fig. 61. Slickensides along underside of bedding plane in Isle la Motte Limestone. Abandoned quarry ¾ mile southeast of Smith's Basin, Washington County.

Chapter 17

Medial Ordovician and the Trenton Tranquility

462–458 million years ago

Subsequent to the crustal unrest of the late Phase II of the Taconian Orogeny, a period of crustal stability ensued that was marked by extensive erosion. The resultant peneplain (tableland) of slight relief became easily overspread by shallow marine water. Within this warm, clear, and mud-free environment, bottom-dwelling invertebrates proliferated. These included abundant and diverse brachiopods, bryozoans, corals, crinoids, and trilobites (Fig. 63). Widespread and numerous limestone strata were deposited within this *"Trenton Sea,"* forming the Trenton Group of carbonate formations. One of the earlier carbonate units to form within this sequence was the relatively thin Balmville Conglomerate and Lime-

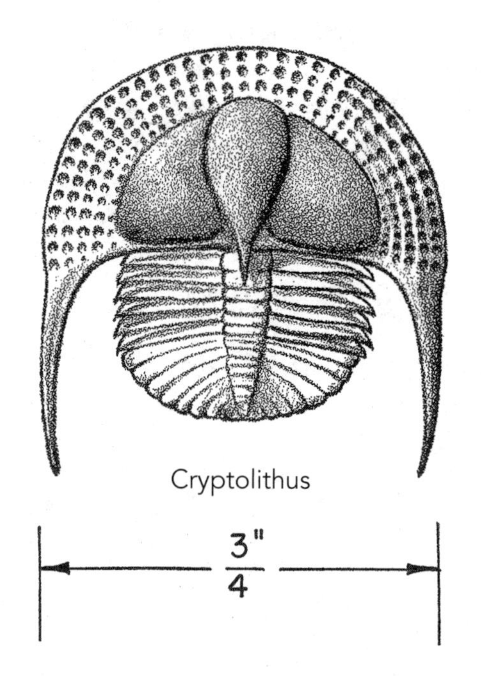

Fig. 63. The trilobite *Cryptolithus tesselatus*, an index fossil to the Medial Ordovician Shoreham Limestone (Upper Glens Falls Limestone), Trenton Group, Mohawk and Champlain Valleys. Printed with permission of the New York State Museum, Albany, NY.

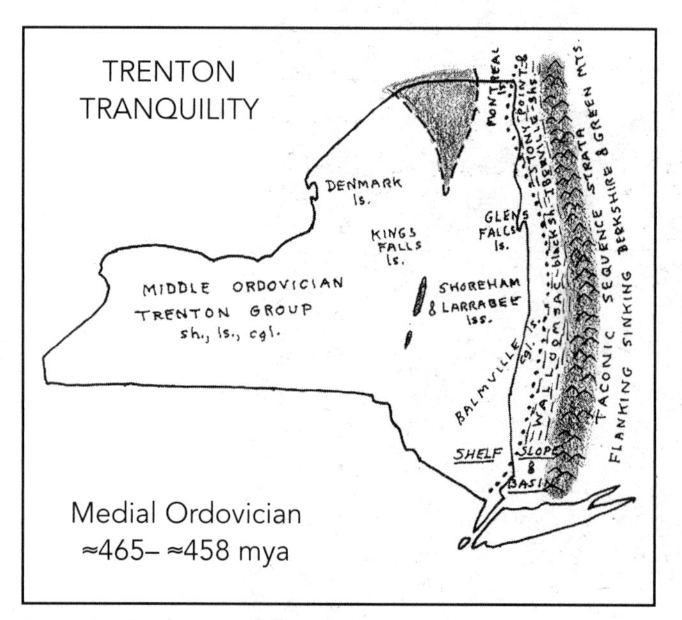

Fig. 62. The Trenton Tranquility.

Fig. 64. Balmville limestone conglomerate, along the east side of Dutchess County 100, 3 miles north of Poughkeepsie.

stone (Fig. 64), ranging in thickness from 0–50 feet. The Balmville Limestone contains a fauna of Early Trenton age. As a result of folding and faulting of the Penobscot Orogeny, the Balmville Limestone rests on differing Cambrian and Lower Ordovician formations of the Wappinger Group.

Locally, knobs of Wappinger Formation rocks projected above the "Trenton Sea," and where the Balmville or other limestones were not deposited, the overlying Walloomsac or Utica black mud would eventually accumulate directly upon the Lower Ordovician Wappinger units.

The Austin Glen Graywacke Puzzle

During the Medial Ordovician a new character appears in the story—the Austin Glen Graywacke. Our understanding of this formation has historically been problematic for three reasons:

1. The determination of its precise stratigraphic position results in contradictory conclusions.
2. Its provenance is unsettled.
3. There is no transition to any shelf carbonates, making the age of the formation ambiguous.

It is the belief of this author that the contradictions and confusion that hinder an understanding of the Austin Glen Formation arise because there are *actually two distinct "Austin Glen Formations"* that were deposited in different places, probably at different times. To best appreciate the arguments to be presented in this section, it is recommended that the simplified Ordovician stratigraphic column, including the "Ordovician Graptolite Zones in New York State," be consulted as an aid to following the discussion.

Fig. 65. Turbidite structures on underside of bedding of Lower–Middle Ordovician Austin Glen Graywacke. (Previously [erroneously!] termed "dinosaur hide" by a local newspaper.) On west side of US 9W, one mile south of Ravena. Exposure shows ripple marks, rill marks, and mud flows.

Fig. 66. Turbidite features in Austin Glen Graywacke, north side of St. Andrews Road, Hyde Park, Ulster County

Let's begin with something we can generally all agree upon—its physical makeup. The Austin Glen Formation can be divided into two members. An upper member consists of massive- to thick-bedded graywackes to subgraywacke rocks, best exhibited at Highland, Ulster County, along US 44 and NY 55 at the western end of the Mid-Hudson Bridge. A lower member of medium- to thin-bedded (and occasionally thick-bedded) graywacke, subgraywacke, and sandstone—all interbedded with gray shale—is widespread in western Columbia and Dutchess counties and eastern Greene and Ulster counties. This lower unit is replete with a variety of sedimentary features that include flute casts (small trough-like depressions or grooves in sediment), ripple and rill marks, graded bedding, and crossbedding. These features are collectively termed **turbidites** and represent a widespread fluid-sediment flow on a soft mud sea floor possessing a greater than normal steepness (gradient). Turbidite features are typical on the outermost continental shelf, and continental slope notched by submarine canyons (Figs. 65, 66, 67).

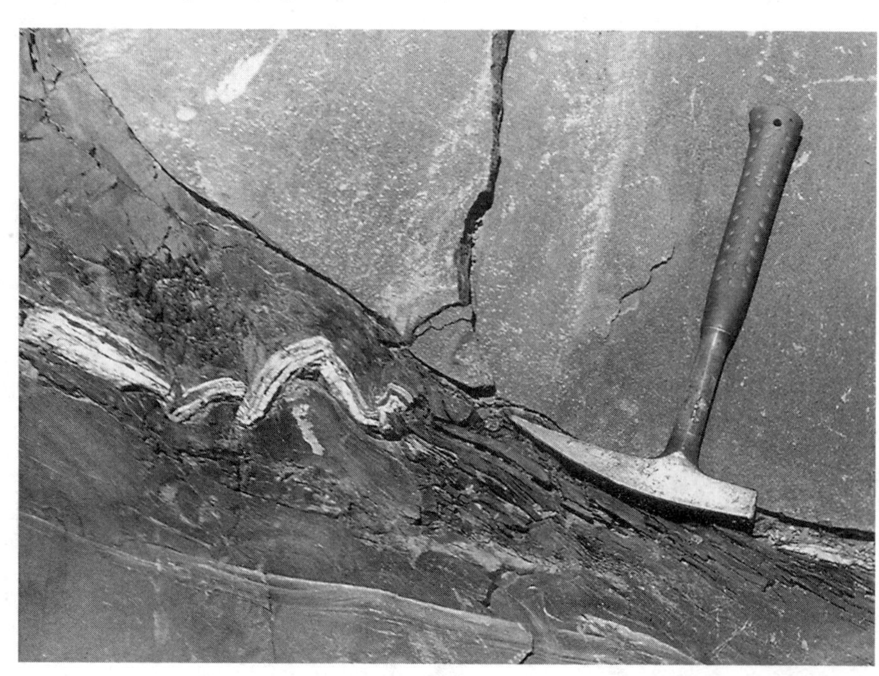

Fig. 67. Crumpled and ruptured vein of white quartz, caused by small thrust fault from the right at the hammer head (graywacke bed over disturbed shale). This demonstrates Taconic fault mechanics on a small scale. Lower Austin Glen Formation, north side of St. Andrews Road, Hyde Park Quadrangle, Dutchess County.

Stratigraphically the Austin Glen lies within the *Taconic Sequence* of clastic strata, which were later transported within thrust slices and gravity slides during Phase III of the Taconian Orogeny. It conformably rests upon the Mt. Merino black shale and black chert. These allochthonous (displaced from their depositional realm) formations outcrop in eastern Greene and Ulster counties and in Washington, Rensselaer, Columbia, and Dutchess counties.

By contrast, Potter (1972) reported the Austin Glen Formation to lie *within* the probably younger Walloomsac-Snake Hill Formation within the *Synclinorium Sequence*, and to be younger than Balmville Limestone. If correct, *this does not correlate to the previously described "Austin Glen."* We will denote this unit as Potter's "westerly Austin Glen." Personally, while mapping in eastern Columbia County (Ancram, Clermont, and Copake Quadrangles), this author discovered Austin Glen lithologies within the Walloomsac slate and stratigraphically younger than the Balmville Limestone that are consistent with Potter's "westerly Austin Glen."

To add to the confusion, Offield (1967) has the Austin Glen below the Snake Hill and above his Mt. Merino shale and, in turn, above the Balmville Limestone. This would seem to be stratigraphically consistent with the originally designated Austin Glen. This author believes, however, that Offield misidentified the Mt. Merino shale. In reality, what Offield believed to be Mt. Merino shale is actually the lowermost Snake Hill Formation (graptolite zone of *Corynoides americanus*); further confirmation of the misidentification is that Offield failed to report the presence of chert, which is ubiquitous in Mt. Merino shale, or its index graptolite, *Nemagraptus gracilis*. When this misidentification is corrected, Offield's placement and dating of the Austin Glen is consistent with Potter's "westerly Austin Glen." Accordingly, the younger "western" Austin Glen must be in the *Corynoides americanus* zone or in the lower *Orthograptus ruedemanni* zone.

Only one conclusion can be drawn: there are *two* Austin Glen Graywackes:

- An *allochthonous* unit deposited east of the Berkshire–Green Mountain Highlands and within the Taconic Sequence of deep oceanic rocks—the "originally recognized" Austin Glen.
- A second, *autochthonous* "western Austin Glen"—located in the probably younger shelf rock Synclinorium Sequence now west of the Berkshire–Green Mountain Highlands metamorphic massif.

Fig. 68. The Austin Glen Graywacke puzzle—solved?

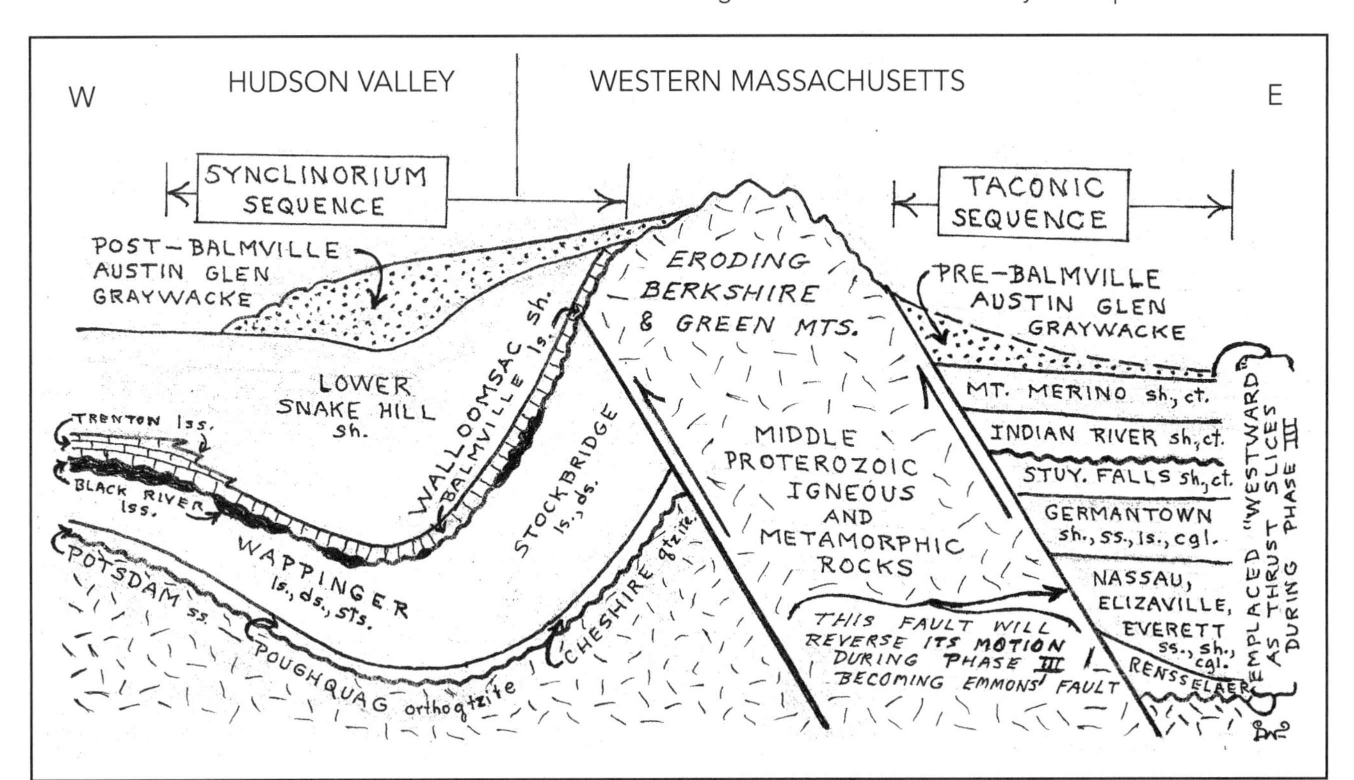

Unfortunately, age correlation of either Austin Glen Formation with age-established, continental-shelf, fossil-bearing limestones cannot be determined because they do not grade laterally into each other. In the past, various workers have equated the Austin Glen, Mt. Merino, and Indian River—members of the Normanskill Group—with either the Chazy, Black River, or Trenton Limestone groups. Because the "western" *Synclinorium Sequence* Austin Glen wedges out within the Snake Hill Shale beneath more recent Silurian and Devonian strata, and above the Early Trenton Balmville Limestone, the "western" Austin Glen is certainly Medial Trenton in age. The "eastern" *Taconic Sequence* Austin Glen is older, based on graptolite evidence.

The recognition of two Austin Glen formations may also resolve a paleontological puzzle. Two graptolite specialists—William B. N. Berry (1962) and John Riva (1974)—proposed differing, and apparently contradictory, graptolite zonation schemes for the New York Ordovician rocks. Both concur that the Indian River and Mt. Merino shales and cherts occupy the *Nemagraptus gracilis* graptolite zone. Above that, their respective zonations differ. Berry regarded the Austin Glen, Snake Hill, and Walloomsac formations as belonging to his *Climacograptus bicornis* zone and, therefore, all of the same age—perhaps a correct interpretation if he was studying the "westerly" Austin Glen. Riva assigned the Austin Glen to his *Diplograptus multidens* zone—a correct interpretation if he was sampling the "eastern" *Taconic Sequence* Austin Glen. The apparent discrepancies between these two respected workers may be resolved by determining in *which* "Austin Glen" their respective collections were made.

A final complicating factor is the issue of formation thickness. In the northern Taconic Mountains of Vermont, the Austin Glen Formation averages 500 feet thick. In the Hoosick Falls area, Potter (1972) reports a thickness ranging from less than 350 feet to over 1000 feet. Near Chatham, Columbia County, the thickness of the Austin Glen is no more than 450 feet, and this author has measured less than 100 feet in the Stottville Quadrangle where it overlies the Mt. Merino black shale. Further to the west—in western Dutchess, eastern Ulster, and Greene counties—the formation thickens to almost 2,500 feet. Sedimentary rock formations are thickest in the direction of the source of sediments; therefore, it would be assumed that the Austin Glen has a "western" origin. This is not possible, however, because to the west the only available source rocks were continental shelf limestones and dolostones of Late Cambrian, Early Ordovician, and Medial Ordovician ages. This carbonate terrane could not furnish the abundant quartz, feldspar, garnet, zircon, and clay minerals that compose "both" Austin Glen Graywackes. At the time of Austin Glen deposition, the sole potential source for these igneous and metamorphic clasts would be the southern Green Mountains and/or the Berkshire Mountains. Therefore, both are older than the Taconic Mountains. Fig. 68 illustrates the author's interpretation of the origin of the two "Austin Glen" formations.

Chapter 18

Medial Ordovician and the Taconian Orogeny (Phase III)[8]

The Arrival of the Taconic Mountains 458–452 million years ago

The Trenton Tranquility was a relatively brief suspension of crustal activity as the Baltican Plate renewed its subduction under the North American Plate. This tectonic compression accelerated the uplift of the Green and Berkshire mountains to Alpine dimensions. To maintain crustal equilibrium, a sinking, trough-like basin developed "west" of the mountain chain—similar to the way one side of a teeter-totter goes up and the other side goes down. This linear trough bordered the parallel high-relief mountains, providing a convenient repository for the now-excessive erosional detritus being produced. Black muds and subordinate silts comprising the Walloomsac, Snake Hill, and early Utica formations began to pile up in this trough, achieving a total thickness of 7,500 feet.

The first two phases of the Taconian Orogeny were preliminary episodes to the initial exposure of the Taconic land mass. Within Phase III—the climax of tectonic plate squeezing—was the far-reaching overthrusting of the entire suite of *Taconic Sequence* formations from "east" of the Green-Berkshire massif over its top, to eventually rest atop the *Synclinorium Sequence*. The structural mechanism of this enormous crustal transportation involved a large recumbent anticline (nappe) with the youngest formations (the Austin Glen and Mt. Merino) at the leading edge. The entire nappe was floored by an extensive low-angle thrust fault—termed Emmons' Fault (or Line). It was originally a high-angle fault bordering the "east" side of the Berkshire-Green Mountains (Fig. 68).

As continued tectonic compression drove the nappe of *Taconic Sequence* rocks "westwardly,"[9] the sequence eventually approached and overrode the Snake Hill mud-filling basin. The leading edge of the nappe had abnormally steep slopes, allowing the unsupported leading edges of various sized clasts ranging from clay and sand to huge boulders to avalanche, by gravity, into the soupy basin as mélange. Represented today by the Poughkeepsie Mélange, it incorporates both large gravity slides from the leading edge of the Livingston and Stottville slices, as well as the torn-up and "bulldozed" limestones and dolostones of the overlapped Synclinorium Sequence.

In Columbia and Dutchess counties this author has been able to recognize and map the two major thrust slices previously mentioned—an older Livingston

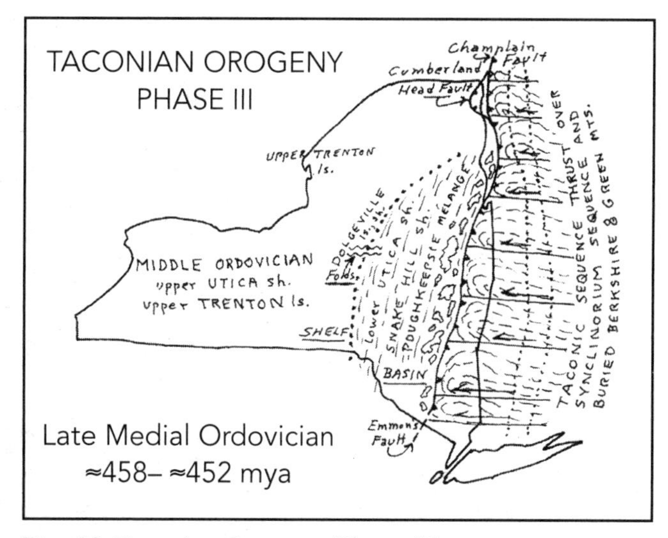

Fig. 69. Taconian Orogeny (Phase III).

Slice and a younger Stottville Slice (Plate 1). Although equivalency has not been proven, the Livingston Slice may equate to the Sunset Lake Slice of Zen (1961). The Stottville Slice may be contiguous with the Giddings Brook Slice of Zen (1961) in west central Vermont and the North Petersburg Sheet (Slice) of Potter (1972) in southern Washington and northern Rensselaer counties. In the Glens Falls–Whitehall area of northern Washington County, the author identified two slices (Fisher, 1984)—an older Hartford Slice and the younger Giddings Brook Slice, together with the associated Forbes Hill Mélange.[10] Similarly, Potter's Whipstock Breccia (1972) satisfies the definition of a mélange. The extensive distribution and unique makeup of the mélange are clues that the "western" margin of the emerging Taconic upland was a precipitous one, marked by thundering spalling-off rock slides and the relicts of tsunami turbulence.

Further evidence of this can be found on I-90 (the New York State Thruway) south-southeast of Little Falls with a large rock exposure that reveals folds in a 1½ meter-thick zone in the Middle Ordovician Dolgeville Formation, sandwiched between unfolded strata. The ribbon limestones and interbedded thin black shales display convolutions varying from symmetrical, to overturned, to broken folds (Fig. 70), and this author concludes that the compressive forces responsible for this unique folding acted at a *specific and limited time*—contorting only a few beds, rather than a thick stratal sequence. The age—*Orthograptus ruedemanni* graptolite zone—numerous fold attitudes, and narrow thickness confinement is compatible with the eastward plummeting of large gravity slides as avalanches into the Snake Hill mud, with attendant seismic events possibly triggering tsunamis in the "Snake Hill Sea." If so, upon reaching the shallow water "west" of the Adirondack Arch (Fig. 60), the tsunami would disrupt the unconsolidated or semi-consolidated strata. For additional details see Fisher (1979).

Fig. 70. Plastically folded Medial Ordovician (uppermost Dolgeville) limestone sandwiched between non-folded earlier Dolgeville below and Utica Shale above. Folding presumably caused by slump and submarine earthquakes, initiated by westward transport of immense gravity slides plummeting into a deepening basin along the meridian of the "Hudson River" during Phase III of the Taconian Orogeny. South of Little Falls, Herkimer County, along north side of New York State Thruway (I-90).

Chapter 19

Late Ordovician and the Taconian Orogeny (Phase IV)

452–445 million years ago

This phase of Taconian mountain building was a continuation of the "teeter-totter" crustal equilibrium condition of rising land accompanying an adjacent sinking basin. The elevated land was the result of the persistent impingement of the Baltica tectonic plate against the Laurentian tectonic plate. In this scene, however, high-angle reverse faults uplifted the earlier far-travelled "Emmons thrust fault" and cross-faulted the nappe so that fault displacement was more vertical than horizontal. To compensate for this upheaval and accelerated erosion, the Snake Hill basin was further deepened and accommodated the voluminous erosional detritus of the Schenectady and Quassaic subgraywackes, sandstones, and gray shales (Fig. 72). Gravity sliding and mélange dumping had now virtually ceased, and these two thick clastic wedges of conglomerate, sand, and silt (with a provenance of an imposing jagged mountain range of the combined Taconic and Berkshire mountains) were incorporated within the mud of the Snake Hill Formation. This coarser detritus reflects the makeup of the newly introduced Curtis Mountain and Rensselaer Plateau Slices. The Schenectady Formation received at least 5,500 feet of sediment, while the Quassaic Group (composed of the Creek Locks, Rifton, Shaupeneak, Slab Sides, and Chodikee formations) has a thickness of approximately 10,000 feet. At this latitude the Snake Hill basin was accumulating erosional debris from the Livingston and Stottville slices. Additional slices have been identified; refer to Plate 1 to locate them. In addition to the Livingston and Stottville slices, the newly emplaced Curtis Mountain and Rensselaer Plateau slices contributed sediment.

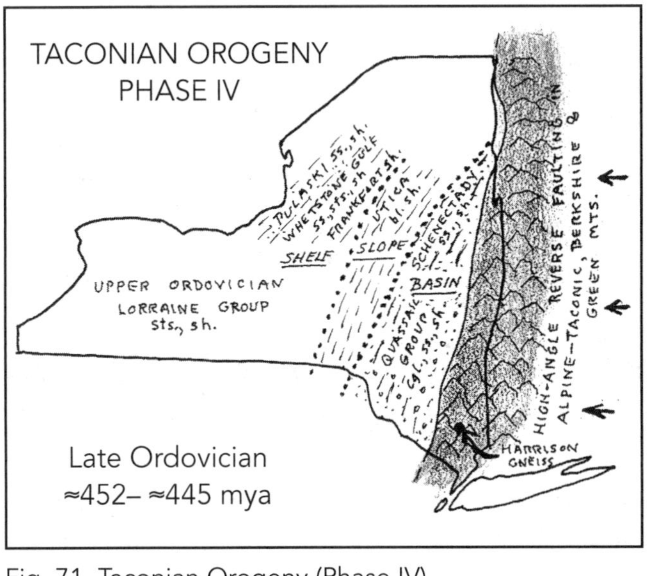

Fig. 71. Taconian Orogeny (Phase IV).

Although the Queechy Lake and the Rensselaer slices may have been connected in the past—based on an identical suite of rocks—they are presently geographically disconnected and overlie different fault slices. The Rensselaer Plateau allochthon rests on the Curtis Mountain allochthon, while the Queechy Lake allochthon rests on the New Lebanon parautochthon. Like the Queechy Lake Slice, the Everett Slice[11] overlies the Synclinorium Sequence formations, but is probably a more "easterly" segment of the Stottville Slice, and thus, the Everett Fault is probably identical to Emmons' Fault, which marks the base of the Stottville Slice.

Fig. 72. Schenectady thin-bedded sandstones and siltstones with shales, along eastbound lane of I-90, south of the West Schenectady Interchange.

In Ulster County this author had recognized the conglomerates, quartzites, and sandstones extending from Hussey Hill on the north, through Shaupeneak and Illinois mountains, to Marlboro Mountain on the south, as being the Quassaic Formation (Fisher, 1971), and estimated the thickness to be at least 3,500 feet. In the Quassaic Formation (Figs. 73, 74), this author observed pebbles of the Austin Glen, Mt. Merino, Indian River, Stuyvesant Falls, Germantown, and Nassau formations. From this observation the author concluded that these sediments were derived from the allochthons of the Taconic Sequence, and not from the Hudson Highlands, since no metamorphic pebbles were found as would be expected from a gneissic Highlands source. This implies that the Highlands *had not yet been emplaced* at the time the Quassaic clastic wedge was forming.

Past presence of the Rensselaer Plateau, Queechy Lake, and Curtis Mountain slices in present-day Dutchess County is uncertain. The relatively large amount of green and gray-green pebbles in the youngest three formations of the Quassaic Group leads one to believe that these younger allochthons, which possessed large amounts of green graywackes, quartzites, and sandstones, probably did exist in Dutchess County.

Fig. 73. Close-up of conglomerate in basal Quassaic Quartzite, west side of railroad one mile southwest of Esopus, Ulster County.

Fig. 74. Quassaic red sedimentary quartzite; ripple marks on underside of overturned beds. Along side of West Shore railroad, one mile southwest of Esopus, Ulster County.

Chapter 20

Late Ordovician and the Taconian Orogeny (Phase V)

445–442 million years ago

During this interval, marine water was absent from what is now New York State. Instead, the area was floored by a peneplain of widespread maroon mud with notably subordinate sand and silt. This iron oxide-rich erosional detritus denotes vigorous erosion from the Taconic, Berkshire, and Green mountains. These thick soils were transported by water "westward" as far as present-day Toronto, Ontario, "northward" to Anticosti Island in the Gulf of St. Lawrence, and "southward" to Virginia. The sediments of this coastal plain formed the youngest Ordovician formation in New York State—the Queenston Shale (tan Oswego Sandstone underlies it). Neither formation has yielded fossils.

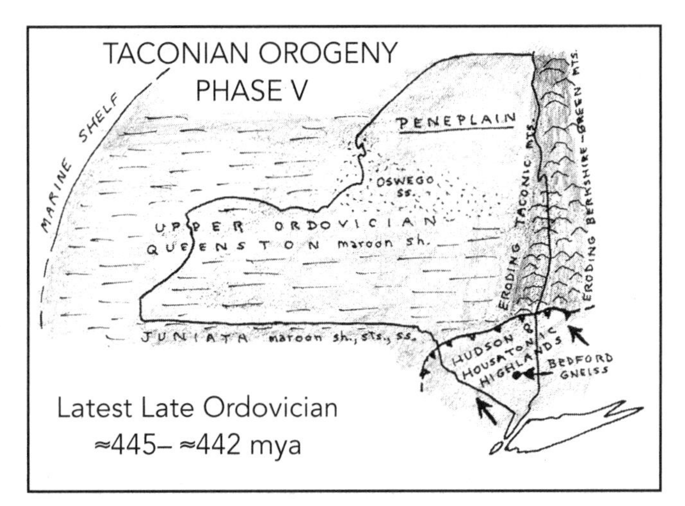

Fig. 75. Taconian Orogeny (Phase V).

It extends into Pennsylvania, where this clastic wedge is termed the Juniata Formation. Here it has a noticeably greater proportion of coarser siltstone and sandstone. This is probably due to the erosion of thicker and more expansive quartz-bearing gneisses of the Hudson Highlands thrust slice emplaced during the close of Phase V compression. This clastic wedge attains a maximum thickness of 4,000 feet and thins to 1,200 feet westward at the longitude of the Niagara River.

Signficantly, in southeastern New York (Rockland, Orange, and Ulster counties) no Queenston maroon shale exists between the overthrust Highland gneisses and the underlying Middle Ordovician Snake Hill Shale, nor between the Early Silurian Shawangunk (pr. shawn-gum) Formation and the Snake Hill Shale. Either the Queenston was not deposited in the Hudson Valley region—which is highly unlikely—or the Hudson–Housatonic Highlands thrust-massif "bulldozed" any maroon strata that had been there. The latter case is more probable, for it explains the abrupt southern termination of the north-south trending Taconic Mountains. Consequently, the conveyance of the northeast-southwest trending Hudson-Housatonic gneisses northwestward[12] was the terminal compressional event of the Taconian Orogeny.

Chapter 21

The Diverse Silurian[13]

The End of an Orogeny, and Environments Anew

442–418 million years ago

The Final Phase of the Taconian Orogeny (Phase V) (442–434 mya)

It may seem surprising to learn that Phase V of the Taconian Orogeny exercised only a slight role in modifying the Taconic Mountains. The only obvious effect is the truncation of the southern end in southern Dutchess County, essentially paralleling Interstate 84. Here, the Hudson Highlands Fault with its contained Middle Proterozoic gneisses has overridden the Wappinger dolostones and limestones of the Synclinorium Sequence. Today, several isolated erosional outliers (Fig. 79) of gneiss atop the Synclinorium Sequence attest to the former geographic extent of the Highlands thrust slice.

The intrusion of ultrabasic igneous rocks of the Cortlandt Complex (hornblendtite, pyroxenite, and emery) now found at Peekskill, Westchester County, finalized the unique crustal alterations of the Taconian Orogeny. This volcanic suite has been radiometrically dated at 434 ± 5 mya, consequently post-Ordovician. It is arbitrary whether this igneous event should demarcate the upper boundary of the Taconian Orogeny or whether it should be placed at the conclusion of the deposition of the Shawangunk-Medina clastic wedge, the erosive material derived from the Hudson-Housatonic gneisses.

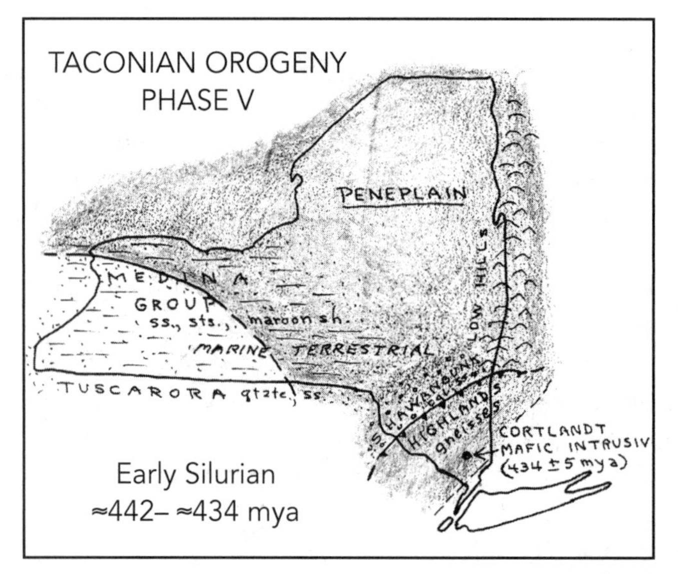

Fig. 76. Taconian Orogeny (Phase V), Early Silurian time.

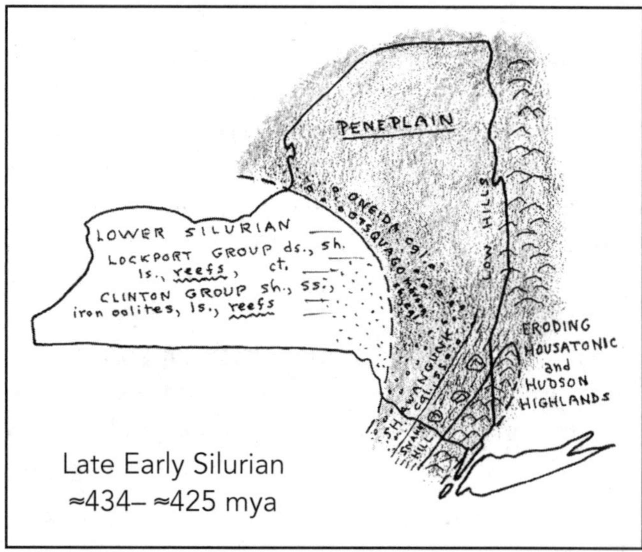

Fig. 77. Paleogeographic map of New York, Late-Early Silurian time.

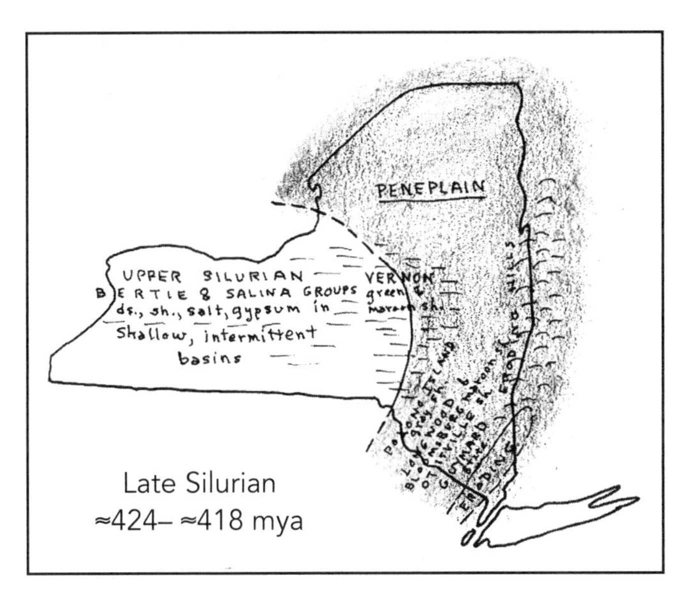

Fig. 78. Paleogeographic map of New York, Late Silurian time.

This resultant agglomeration of gravels, sands, and silts accumulated to a thickness of approximately 2,500 feet from Kingston southwestward to Port Jervis. This imposing hogback, paralleling U.S. 209, comprises the Shawangunk Mountains and continues in New Jersey and Pennsylvania as the Kittatiny Mountains. Its uniformly northwest-dipping strata include the Shawangunk quartz-pebble conglomerate and quartz sandstone, Guymard quartzitic sandstone, Otisville sandy shales, and Wurtsboro sandstones capped by Bloomsburg and Longwood maroon shales and maroon sandstones. None of these formations has yielded marine

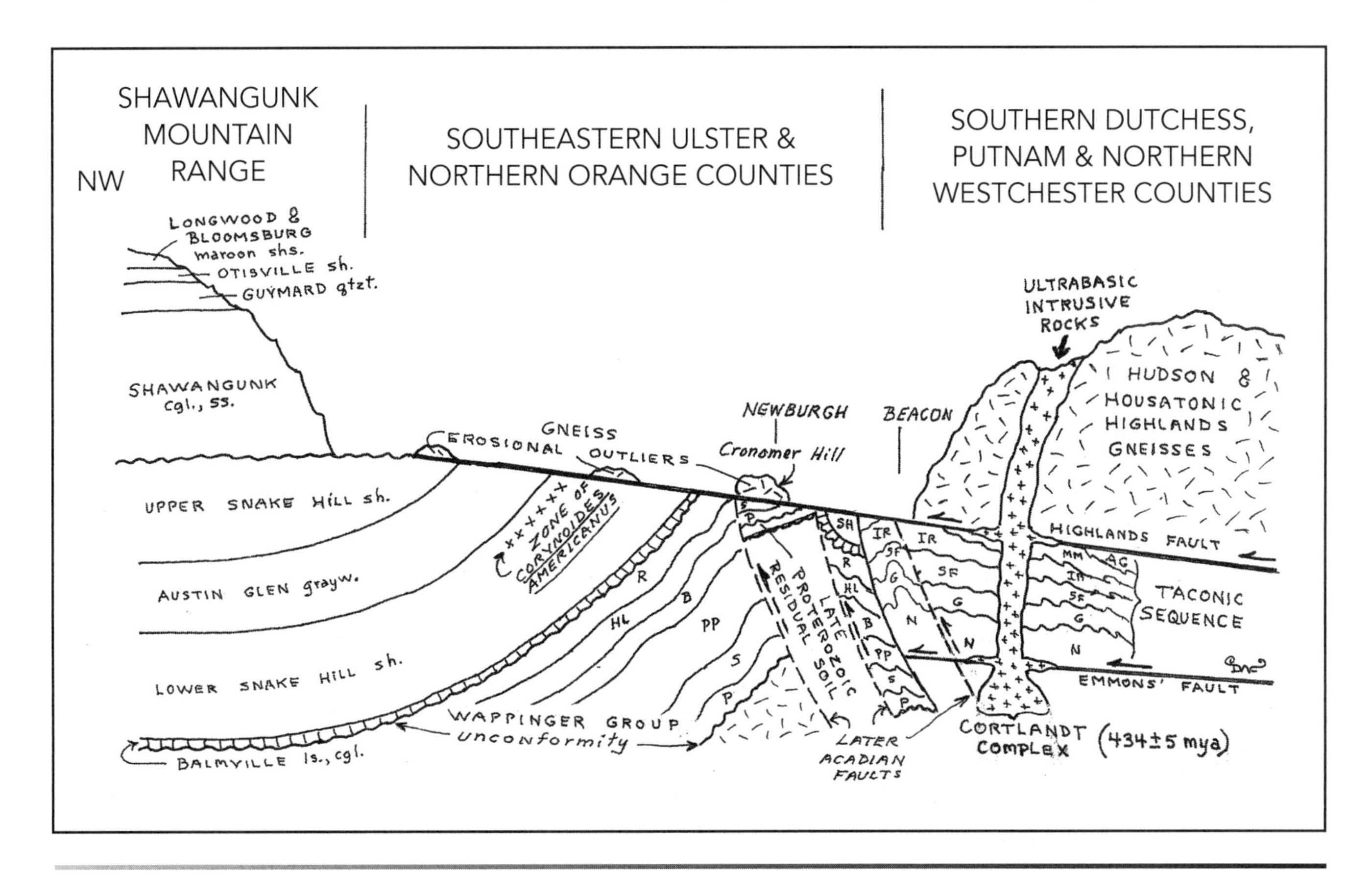

Fig. 79. Taconian Orogeny (Phase V)—Early Silurian.

fossils; hence they are considered to be terrestrial in origin. When traced westwardly into the subsurface, these formations grade laterally into the Medina and Clinton clastic strata. The Shawangunk ridge and its formations extend to the north as far as 9 miles southwest of Kingston. Exceedingly meager clastic material was available as the Taconian thrust slices undoubtedly were beveled to near sea level. From the Kingston area north, Upper Silurian thin limestones and dolostones rest unconformably on Middle Ordovician graywackes, shale, and mélange.

Meanwhile, south of Kingston, cascading mountain streams tumbling off the newly emplaced Hudson Highland gneisses carried gravel, sand, and silt "northwestward" as braided streams meandered across a narrow tableland floored by Ordovician strata. A virtual absence of clay minerals (except for those found in the youngest Bloomsburg and Longwood maroon shales) attests to derivation of the remaining Early Silurian formations from the Middle Proterozoic rocks.

Summary: Some Generalizations about the Taconian Orogenesis

1. During the earlier portion of the Taconian Orogeny (470–458 mya), crustal instability events (Phases I and II) were shorter and alternated with longer crustal stability episodes, termed *tranquilities* and represented by the formation of Chazy, Black River, and Trenton limestone suites.
2. Formation ages and stratigraphic successions of allochthons of Phases III and IV reveal that younger formations were emplaced earlier or at the leading edge of thrust slices, whereas older formations occupied the trailing segment of the slices or were transported later.
3. Thickest and most extensive erosional detritus (clastic wedges) corresponds to the phases exhibiting continual uplift and highest relief.
4. Focus of greatest intensity of crustal compression generally migrated "southward" during the five phases of the Taconian Orogeny.

The Diverse Environments of Silurian Seas

Because the Silurian Period is a relatively short interval (24 my) as compared to other Paleozoic periods, it is currently—and arbitrarily—divided into only Early or Late divisions. In New York State, the Early Silurian is designated the Niagaran Series from the continuous sequence exposed in the Niagara River Gorge along the New York–Ontario border. The Late Silurian is designated the Cayugan Series from the many exposures north of the Finger Lakes region in central New York in Wayne, Cayuga, and Onondaga counties. The early portion of the Niagaran Series, including the Medina Group, has been discussed in the preceding section concerning the late phase of the Taconian Orogeny. In New York there are no Silurian rocks east of the Hudson River.

The remaining ≈ 11 my of the Niagaran Series—represented by the Clinton and Lockport Groups—demonstrate a period of crustal stability (Fig. 77). Vastly different types of short-duration environments are represented. The Clinton Group ranges in thickness from 120–330 feet and begins with a basal, unfossiliferous white sandstone possibly terrestrial in origin, designated the Thorold Sandstone. It is followed by a quartz-pebble conglomerate (the Oneida Conglomerate) changing upward into olive-green spore-bearing shales (the Neahga and Maplewood Shales). Next upward are very fossiliferous reef-bearing limestones (the Reynales and Irondequoit limestones) and local incongruous graptolite-bearing black

Williamson Shale. Atop the Irondequoit Limestone is the very fossiliferous Rochester calcareous gray shale. South of Utica, clastic formations different in color and texture prevail—the Sauquoit, Otsquago, Willowvale, and Herkimer formations. The Otsquago is strongly crossbedded. Unique to some Clinton formations are thin, one to four-inch layers of *oolitic* hematite—an aggregate of tiny sphere-shaped particles. Oolitic hematite probably represents runoff from adjacent soil.

Atop the Rochester Shale is the Lockport Group of locally reef-bearing fossiliferous limestones and dolostones—the caprock of Niagara Falls. It has a thickness ranging from 100–180 feet. The Lockport Group is an important economic resource in western New York State, supplying crushed limestone and dolostone for concrete aggregate and other construction uses. In addition, the formation contains pockets with well-developed crystals of minerals that include fluorite, dolomite, calcite (Fig. 80), sphalerite, gypsum, and galena—highly sought-after by mineral collectors. Neither the Clinton nor Lockport groups extends east of southern Herkimer County.

Environmentally, the Cayugan Series is the antithesis of the Niagaran. These next ≈7 million years record an arid landscape with basins of intermittent highly saline water hosting an occasional sparse fauna. Rare, environmentally dubious clams and snails have been found in halite (rock salt) and gypsum-bearing shales of both the Salina and Bertie groups (Fig. 78); these formations have thicknesses of 50–1,200 feet. Bryozoan-coral reefs, trilobites, crinoids, and plentiful brachiopods—all characteristic of the earlier Lockport and Clinton groups—are notably absent. Clearly, access to normal-salinity marine seas was unavailable. Mohawk Valley normal faults may have been active at this time, producing a barrier (elevated topography) to connection with normal marine water. A peculiar faunal exception was a puzzling group of arachnid arthropods—eurypterids (u-rip′-tur-ids). While globally they are exceedingly rare, these "sea-scorpions" evidently led a euphoric existence in the brine-rich Late Silurian waters in present New York, and especially within the Bertie inland sea (Fig. 81). Extraordinary fossils of these animals have been collected in a belt of Cayugan Series strata extending from south of Utica westward to

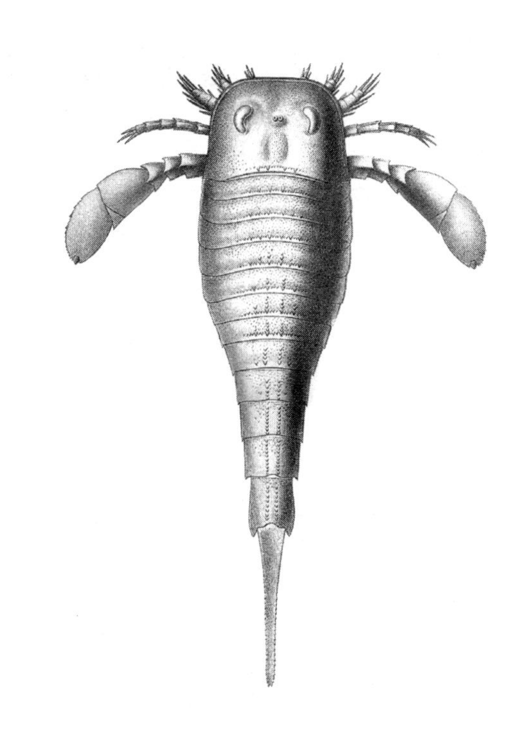

Fig. 81. Illustration of *Eurypterus remipes*, the New York State fossil, from Clarke and Ruedemann, *The Eurypterida of New York* (1912). Actual length of fossil is 9 inches (22 cm). Some eurypterids grew to several feet in length. The stratigraphic horizon and geographic location is described as "Bertie (siluric) waterlime, Herkimer County"; current usage would describe the fossil as being found in the Bertie Dolostone of Silurian age.

Fig. 80. A group of calcite crystals from Lockport, Monroe County. The field of view is approximately 3 inches (8 cm). Photograph by Steve Nightingale.

Fig. 82. Fossil of a juvenile *Eurypterus remipes* collected by author during widening of the "Passage Gulf" road cut near Spinnerville in southern Herkimer County. The relatively high concentration of eurypterid fossils found at this location suggests that this locality represents a possible breeding pool for these animals. Photograph by Steve Nightingale.

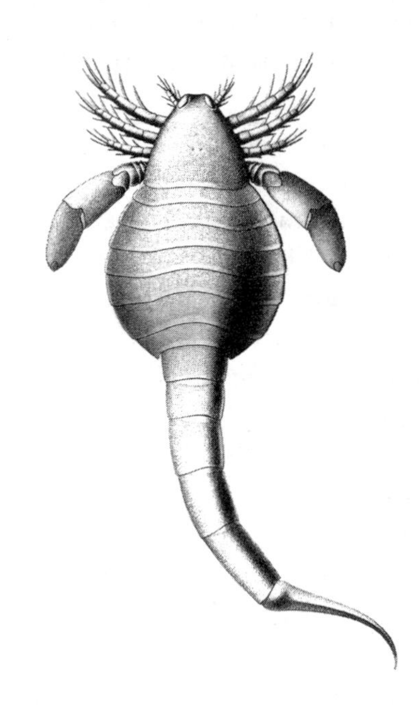

Fig. 83. Illustration of *Eusarcus scorpionis*, from Clarke and Ruedemann, *The Eurypterida of New York* (1912). Size approximately 10 inches (25 cm).

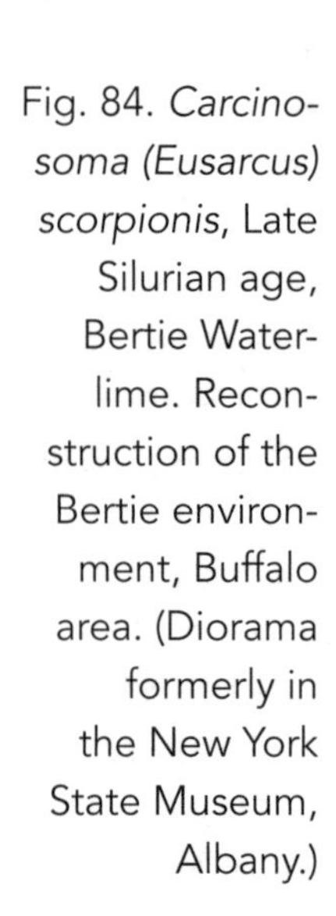

Fig. 84. *Carcinosoma (Eusarcus) scorpionis*, Late Silurian age, Bertie Waterlime. Reconstruction of the Bertie environment, Buffalo area. (Diorama formerly in the New York State Museum, Albany.)

Buffalo, resulting in its appropriate designation as the New York State Fossil. Of evolutionary significance is the fact that the oldest known land scorpions have been recovered from the Bertie Dolostone.

Within the last few hundred thousand years of Silurian time, the hypersaline waters became diluted by the breaching of marine waters of more normal salinity. This caused an influx of more typical marine faunas. In the western Hudson Valley, within the Rondout Formation (thickness 0–75 feet), are the basal Wilbur-Cobleskill limestones (Fig. 85) supporting a cosmopolitan marine fauna including corals. The Rondout Formation also includes the Rosendale Dolostone (a natural cement rock; see "Economic Resources from the Ground in Columbia County") and the Glasco Limestone, which in places is a solid 5-foot-thick blanket reef of "chain-link" corals termed *Halysites* (Fig. 86)—an index fossil for

Fig. 85. Upper Silurian Wilbur Limestone resting on Middle Ordovician Austin Glen Graywacke (angular unconformity). West side of NY 32 at north edge of Kingston, Ulster County.

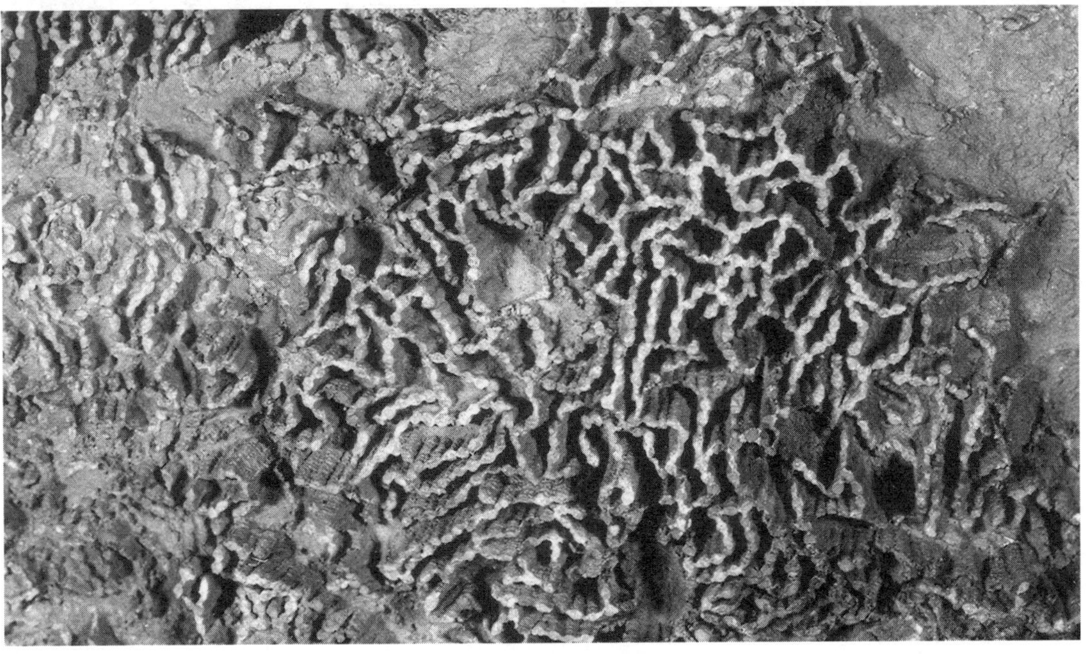

Fig. 86. Colonial "chain" coral *Halysites*, index fossil to the Silurian Period; field of view approximately 6 inches. Specimen from Late Silurian Glasco Limestone member of the Rondout Formation, NY 32 on northeast edge of Kingston, Ulster County.

the Silurian Period. The uppermost member of the Rondout Formation is the Whiteport Dolostone—another natural cement rock. Rickard (1975) regards it as having an Early Devonian age; thus, this formation has the unusual distinction of spanning two geologic periods! Mudcracks (Fig. 87) in the Rosendale and Whiteport Dolostones are produced by alternate wetting and drying in the supratidal and intertidal zones—demonstrating the beach environment of the Rondout Formation Dolostones and the shallow subtidal zones of the Rondout Limestones. Scarcity of sand and silt is noteworthy.

Silurian lands remained largely barren of life, though some primitive types were undoubtedly present as suggested by trace fossil patterns in nonmarine strata. Primitive land plants necessarily preceded any encroachment of animals. In New York, during deposition of the Bertie Formation, rudimentary plants had appeared. These immigrants from the sea were the vanguard for the army of animals that were to invade the land in the Devonian Period.

Fig. 87. Mud (desiccation) cracks in dolostone demonstrating intertidal origin. Upper Silurian Rosendale Waterlime (Dolostone) of Rondout Formation, west side of NY 32, north edge of Kingston, Ulster County.

The Taconic Resurrection: The Devonian[14] Renaissance and the Acadian Orogeny (418–362 million years ago)

"The interest in a science such as geology must consist in the ability of making dead deposits represent living scenes."

Hugh Miller[15] (1802–1856)

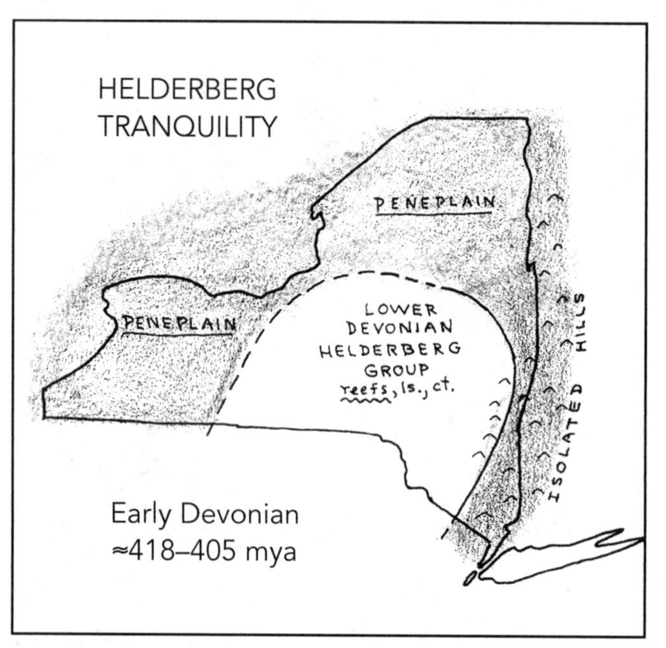

Fig. 88. Paleogeographic map of New York.

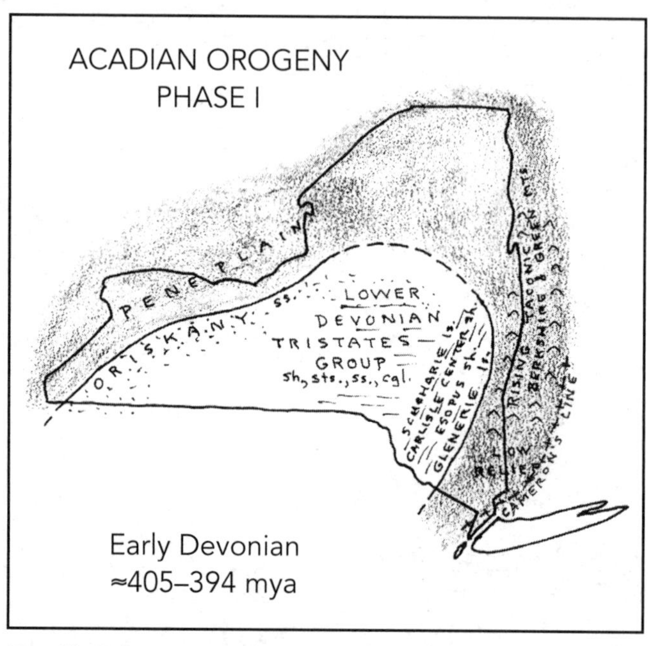

Fig. 89. Paleogeographic map of New York, Acadian Orogeny (Phase I).

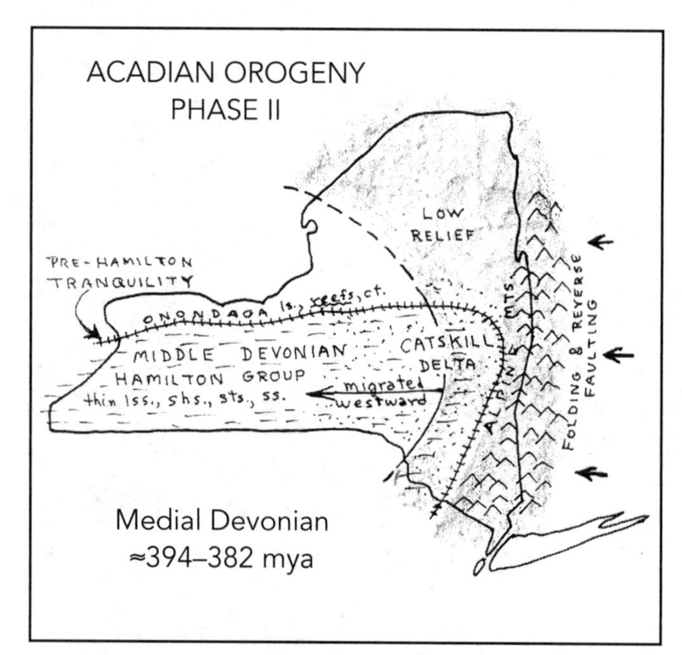

Fig. 90. Paleogeographic map of New York, Acadian Orogeny (Phase II).

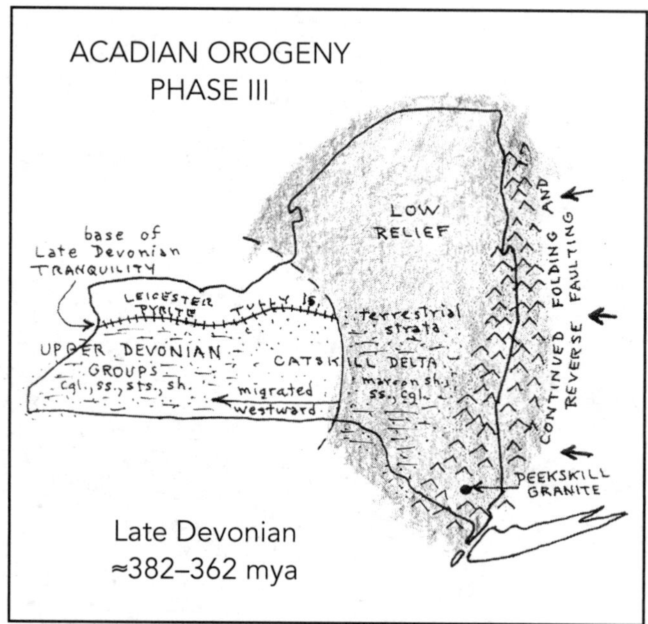

Fig. 91. Paleogeographic map of New York, Acadian Orogeny (Phase III).

Chapter 22

Early Devonian

418–394 million years ago

The curtain on the Devonian age rose on a scene similar, but not identical, to the youngest Silurian one in which the Rondout Formation formed—a situation where the Taconic Mountains were considerably lowered from their former stature. The "Low Taconics" featured a flat coastal plain with an irregular shoreline breached by shallow bays interspersed with rocky peninsulas. An archipelago of craggy, upturned rock islands functioned as partial barriers to waves and longshore currents, thereby protecting backwater lagoons. Acting in this capacity were, among others, Blue Hill and Mt. Merino in Columbia County and Hussey Hill, Illinois Mountain, and Shaupeneak Mountain in Ulster County. Where offshore islands did not obstruct access to the open ocean, barrier blanket reefs of cabbage-like stromatoporoids (an extinct group of laminated, sponge-like colonial animals) (Fig. 92) served as wave breakers.

The "High Taconics" were a terrane of low ridges and hogbacks with very low gradient streams draining westward into the Early Devonian *Helderbergian Sea,* so named for the group of carbonate rocks that formed there. During this *Helderberg Tranquility,* initial Devonian flooding encroached upon a stable, relatively flat land that was supplying exceedingly little clastic material to the sedimentary realm. The Helderberg suite of limestones that formed in this environment is splendidly displayed in a prominent escarpment paralleling I-87 west of the Hudson River in Albany, Greene, and northern Ulster counties. It is less well exposed southwesterly from south of Kingston to Port Jervis on the New Jersey border.

Fig. 92. Cross section of lumpy, coral-like reef of extinct stromatoporoids, indicating warm, shallow, clear marine water. Early Devonian Manlius Limestone, field exposure south of town road 1.35 miles northwest of Sharon, Schoharie County.

Fig. 93. Bedding plane surface filled with an extinct class of "screw-shaped" mollusks (Cricoconarids or Tentaculids), *Tentaculites gyracanthus.* Lower Devonian Manlius Limestone, outcrop on south side of town road 1.4 miles northwest of Sharon, Schoharie County.

The basal member of the Helderberg Group is the Manlius Limestone. This is a fine-textured, dark gray to black rock formed in intertidal and lagoonal environments—exposed today westward from Thacher State Park[16] and extending to Union Springs in Cayuga County. Sedimentary clues to a sometimes intertidal origin are mud- or sun-cracked strata and thin beds of cross-laminated, coarse-textured limestones. The Manlius fauna is atypical, as it is characterized by few and diminutive species, but immense numbers of them. Certain bedding planes are crowded with either the needle-like, extinct molluscan tentacultids (*Tentaculites gyracanthus*) (Fig. 93), the bean-shaped ostracode (*Leperditia alta*) or the small, strongly ribbed spiriferid brachiopod (*Howellella vanuxemi*). East of Sharon Springs, Schoharie County, the author discovered two bedding planes replete with exceedingly rare fossil edrioasteroid echinoderms (Fig. 94)—the class is now extinct.

Fig. 94. Bedding planes of extinct edrioasteroid echinoderms. Note five arms on disc about the size of a 5-cent coin. Early Devonian Manlius Limestone, road cut on south side of town road, 1.35 miles northwest of Sharon, Schoharie County.

Age-equivalent to the quiet-water, low-energy Manlius Limestone is the Coeymans Limestone, which formed in a very different environment. The Coeymans Limestone formed in a shallow, subtidal, rough-water, high-energy realm. This limestone is characterized by a coarse texture, light gray color, and contains abundant fossilized horn-shaped solitary corals and pie-shaped colonial reef corals. Both denote clear, warm, shallow, agitated water of the subtidal zone. Thick-shelled, highly biconvex, coarsely ribbed brachiopods (*Gypidula coeymanensis, Uncinulus mutabilis, Atrypa reticularsis*) likewise attest to the rough-water environment—the surf zone—as do the prolific haphazard array of "lifesaver"-shaped stems (columnals) of crinoids. The overlying Kalkberg black chert-bearing limestone and its superjacent New Scotland clayey limestone and interbedded shale document progressively deeper water below the level of wave base. Both formations are dominated by bryozoans and flatter brachiopods (Fig. 7), indicating quieter water on the more distant continental shelf.

The younger Becraft–Alsen–Port Ewen limestones are an analogous cycle of the Coeymans–Kalkberg–New Scotland sequence. Particularly interesting is the coarse-textured, fossil-fragmental Becraft Limestone. This buildup of crinoidal debris, principally columnals, is interwoven with hemispherical crinoid bases or "Scutellas" (*Aspidocrinus scutelliformis*), together with strongly biconvex uncinulid and atrypid brachiopods. Like the Coeymans Limestone, the Becraft Limestone documents an above-wave base, high-energy environment. Unlike the nearshore Coeymans, the Becraft was a further offshore sea-mount submarine tableland crowded with crinoid "meadows" (Fig. 95). These crinoid thickets may have been so impenetrable that they may have discouraged mobile dwellers; this may explain the virtual absence of swimmers in the Becraft habitat. Whereas the Alsen cherty limestone is a Kalkberg "look-alike," the Port Ewen differs lithically and faunally from the New Scotland. An instructive exposure along NY 199 at the northern edge of Kingston demonstrates these differences. Here, football-sized limestone nodules occur throughout a calcareous mudstone matrix (Fig. 96). The Port Ewen is unknown north of Green's Lake (north of Catskill), Greene County. Southward, the formation thickens. The relative thicknesses of the Helderberg Group formations are illustrated in Fig. 97.

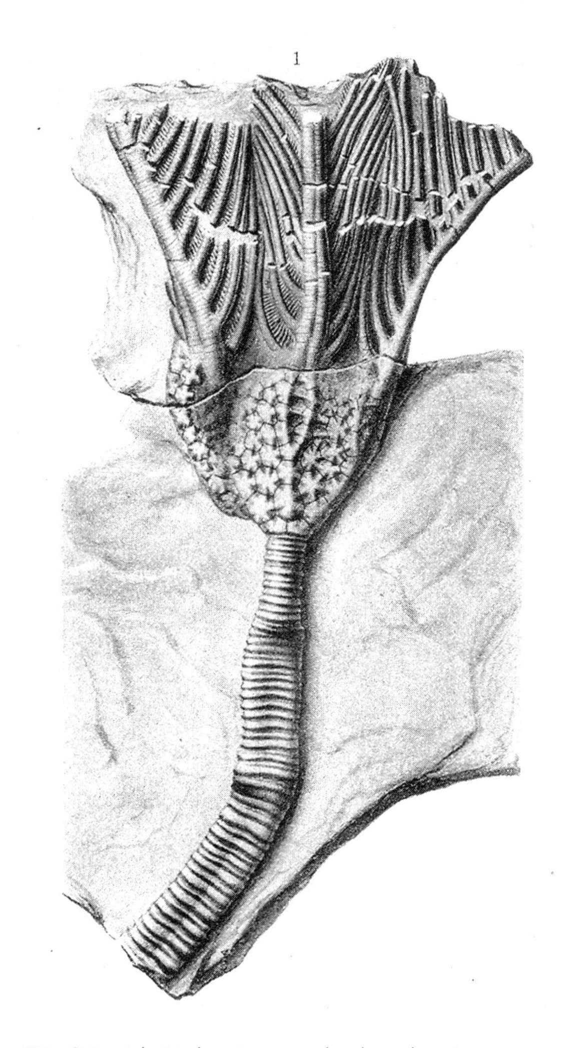

Fig. 95. Crinoid, *Melocrinus pachydactylus*, Lower Devonian Coeymans Limestone. From Goldring, *The Devonian Crinoids of New York State*, 1923.

In New York, the Helderberg limestones represent at least seven different environments and rank high in variety and abundance of invertebrate fossils. Bottom-dwelling filter-feeders dominate, foremost of which are the varied brachiopods. These have a convexity ranging from nearly spherical (for those species living in rough water) to nearly flat (for those living in quiet water), and shell surface grading from coarsely ribbed to smooth. Secondary filter-feeders are the bryozoans and corals; the presence of these animals signifies an environment of clear, warm, shallow, marine water atop a stable crust. Completing the Helderbergian faunas were sponges, high- and low-spired snails, nautiloid cephalopods, crinoids, rare clams, and extinct cystoids and trilobites. Judging from the cluttered graveyards of invertebrate fossils in these limestones, the Helderberg Sea was a flourishing

Fig. 96. Limestone nodules in Early Devonian Port Ewen (Helderberg Group) calcareous mudstone, north side of NY 199 at north edge of Kingston, Ulster County.

sanctuary, free from vagabond predators. But an innocuous, one-inch volcanic ash bed (radiometrically dated at 395 ± 5my by Donald Miller at the Rensselaer Polytechnic Institute) within the Kalkberg Limestone—an omen of impending crustal unrest—signifies that there was trouble looming on the horizon (Fig. 98).

Whereas Helderberg limestones are virtually free of quartz-sand and silt, with minimum clay content, the overlying Tristates Group is carbonate-poor and sand-shale rich.

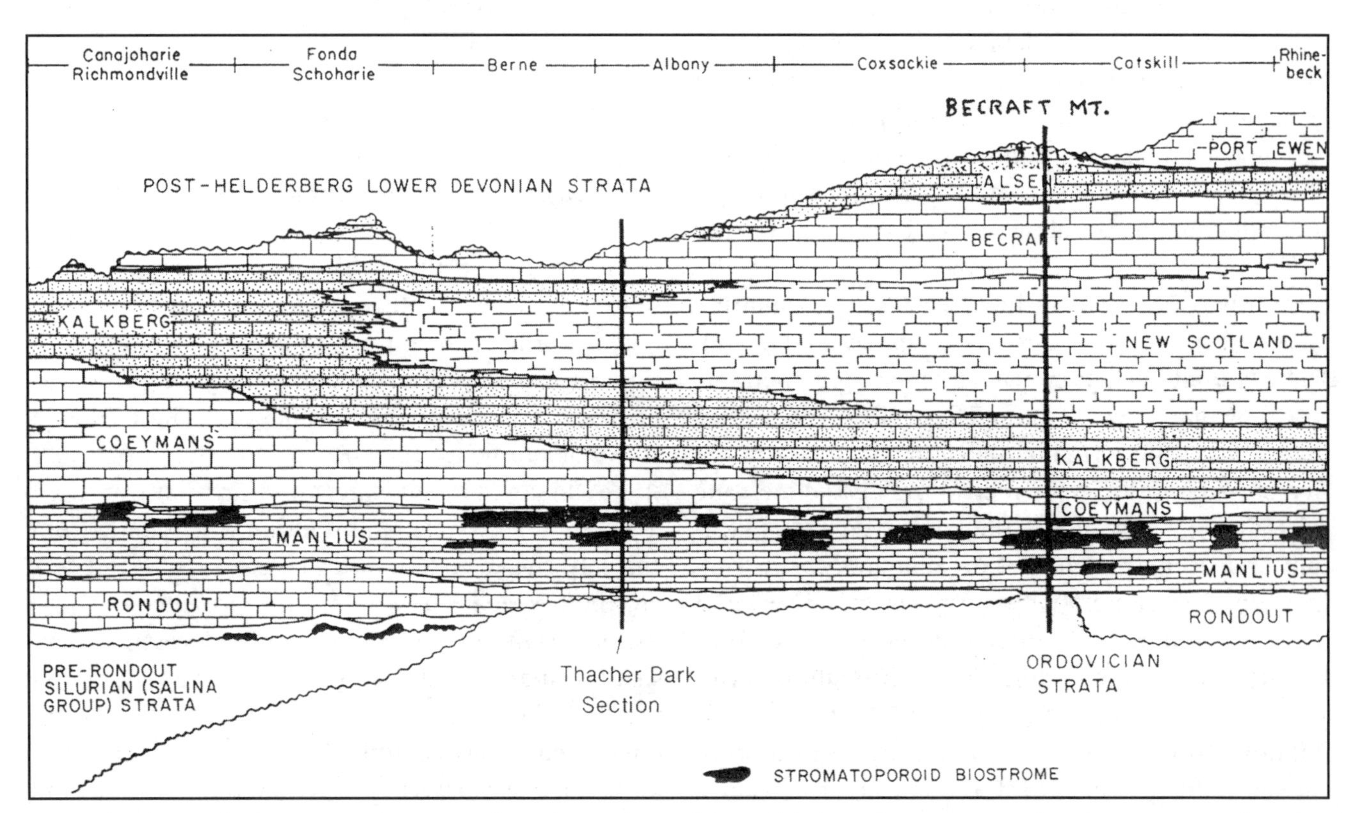

Fig. 97. Relative thicknesses and stratigraphic relationships of the Helderberg limestones in the 15-minute quadrangles referred to in the text.

These clastic formations, from oldest to youngest, are: the Connelly Conglomerate (spotty and limited in extent; Oriskany Sandstone (equivalent to the siliceous and clayey Glenerie Limestone); Esopus sandy and silty shale and chert; Carlisle Center clayey-calcareous siltstone; and Schoharie calcareous mudstone. The interpretation of this initial Devonian erosion-derived material is quite obvious; a marked change of sedimentary residue signifies renewed crustal instability. Here, the crustal instability was accompanied by volcanic activity, as evidenced by layers of volcanic ash found in the Tristates Group.

The Taconic Mountains, having been eroded to a low surface of little relief, were about to stage a comeback in the later Early Devonian. The author regards this tectonic episode as Phase I of the *Acadian Orogeny* (Fig. 89). This orogeny was the result of a rogue microtectonic plate, termed *Avalonia,* sideswiping Laurentia along a major shear zone—probably represented by Cameron's Line[17]—reaching from Newfoundland, through the Maritime Provinces of Canada, and extending south to Virginia and the Carolinas. Throughout the Devonian Period, this crustal cataclysm took place within a continuously constricting Iapetus Ocean.

Although the same groups of invertebrates continued to be found in Tristates seas, populations of differing formations bear little resemblance to one another. Undoubtedly, factors such as volume, type, and persistence of clastic influx, as well as water turbidity and rigidity of the sea floor, played primary roles in the presence or absence of particular faunas. The basal Oriskany Sandstone or sedimentary quartzite rests atop the Helderberg Group or Upper Silurian strata on an erosional surface resulting from differential uplift. For reasons unknown, the Oriskany's species are abnormally large and include: brachiopods (*Costispirifer arenosus, Hipparionyx proximus, and Rensselaeria ovoides*); fragmentary trilobites (*Dipleura and Synphoria*); a clam (*Actinopteria textilis*); platyceratid gastropods; and nautiloid cephalopods. Primarily composed of well-rounded, frosted quartz sand with interstitial limonite, detrital garnet, biotite, amphibole, and pyroxene, the formation's mineral suite

Fig. 98. Re-entrant (at hammer head) of volcanic ash bed in Kalkberg Limestone, south side of US 20, north of Cherry Valley, Otsego County. Photograph by Steve Nightingale.

Fig. 99. Close-up of bedding plane surface of Carlisle Center Formation (Early Devonian) showing feeding trails of the marine worm *Taonurus* (or *Spirophyton*) *caudi-galli*. East side of NY 166, 1.7 mi north-northeast of Cherry Valley, Otsego County.

suggests derivation from a metamorphic terrane—probably the southern Canadian Shield.

Except for rare small brachiopods (*Leptocoelia and Orbiculoidea*) in the Lower Esopus clayey siltstones and cherts, the remainder of the Esopus and overlying Carlisle Center formations have yielded no fossils other than curious, overlapping, arc-shaped markings termed *Taonurus* (or *Spirophyton*) *caudi-galli* (Fig. 99). These superficially resemble "rooster's tails," and they may represent feeding trails of worms; cross-sections of these features show black carbonaceous nested chevrons, possibly the remnants of ingested material. The sedimentary realm of the *Taonurus* strata is puzzling; possibly it was an elongate, narrow basin. The overlying Carlisle Center Formation is permeated with dull green to bluish green glauconite, a mineral formerly thought to indicate intertidal or supratidal zones. Today, this idea is controversial. Lack of typical marine fossils such as brachiopods, mollusks, and corals in the *Taonurus* zone suggests, perhaps, a terrestrial environment. Equally plausible is a deeper sea realm with densely muddied waters unable to accommodate floaters and swimmers, and a soupy-soft sea floor preventing bottom dwellers from anchoring on a firm substrate. In contrast, the overlying Schoharie calcareous mudstone fauna is typically marine, with brachiopods, solitary corals, and bryozoans dominant—though all species are relatively small except for scarce nautiloid cephalopods.

The Tristates Group is absent west of southern Herkimer County, save for subsurface extension of the Oriskany Sandstone into southwestern New York and its spotty surface outcroppings westward through the Finger Lakes area. Even though Helderberg and Tristates formations successively wedge out westward from the Hudson Valley, no evidence has been found that they were folded or thrust-faulted prior to Onondaga Limestone deposition. Their absence in central and western New York may indicate original westward extension and later removal by pre-Onondaga erosion. Conceivably, there was no Early Devonian sedimentation west of central New York because of regional uplift in the Taconics during Phase I of the Acadian Orogeny. West of Cayuga Lake, the 390 million-year-old Middle Devonian Onondaga Limestone rests on the erosional surface of the Upper Silurian Akron (equivalent to Cobleskill) or Bertie dolostones.

Chapter 23

Medial Devonian

394–382 million years ago

Medial Devonian time commenced with the statewide deposition of the Onondaga Limestone. This very fossiliferous, coral reef-bearing extensive blanket heralds the return of crustal stability—the *Onondaga Tranquility*. The Onondaga Limestone reaches as far west as the Detroit area. It is the youngest limestone found in the Hudson Valley and contains tan chert in its lowest member, and black layered and nodular chert (Fig. 100) in its upper portion. Both colonial and solitary corals predominate, with brachiopods a close second. Bryozoans, gastropods, straight and coiled nautiloid cephalopods, and trilobites (Figs. 101, 102) are subordinate. Occasional ostracodes and fish remains may be recovered. Locally, a basal sandstone or siltstone may be found. Onondaga's serene environment was to be short-lived. Several volcanic ash layers have been discovered within the middle and upper Onondaga Formation, and frequent and widespread volcanism forecasts renewed tectonic activity. One 6-inch layer, six feet from the top of the Onondaga along U.S. 20 at Cherry Valley, Otsego County, thickens to 30 inches in West Virginia. The ash has been radiometrically dated to have an age of 390 ± 0.5 my.

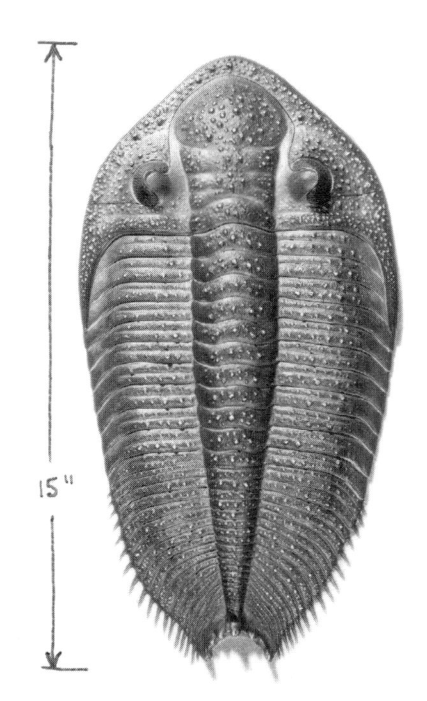

Fig. 101. Trilobite covered with spinose tubercles (*Coronura myrmecophorous*), a crawler on the sea floor. Early Medial Devonian Onondaga Limestone, Kingston, Ulster County. Printed with permission of the New York State Museum, Albany, NY.

Fig. 100. Nodular chert in Middle Devonian Onondaga Limestone, North Kingston, Ulster County.

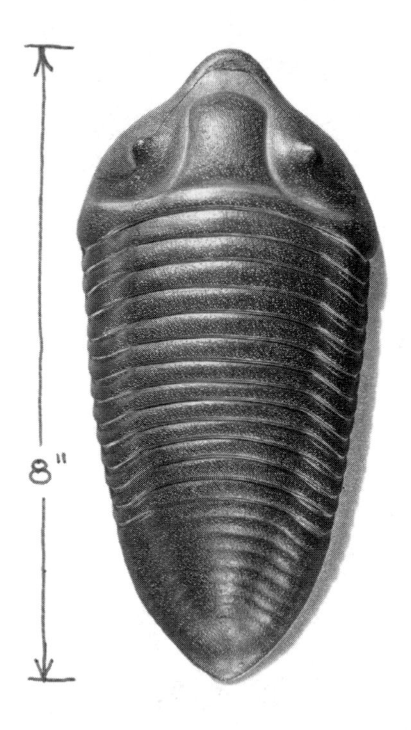

Fig. 102. Trilobite, smooth surface (*Dipleura dekayi*), a burrower into the sea floor. Medial Devonian, Hamilton Group (siltstones, silty shales), Otsego Lake to Cayuga Lake, Madison County. Printed with permission of the New York State Museum, Albany, NY.

The resurrection of the Taconic Mountains accelerated with the geographically spacious and violent *Phase II of the Acadian Orogeny.* This involved westward high-angle reverse faulting, isoclinal folding, and folding of regional Taconian cleavage (Fig. 5). These structural rearrangements extend to a few miles west of the present Hudson River and are well displayed along NY 23 north of Catskill. Devonian-age granite intrusions are conspicuous in Maine and New Hampshire. In New York State, a knife-edge abrupt sedimentary change exists at the summit of the Onondaga Limestone with overlying black shales—the Bakoven and Union Springs members of the Marcellus Formation of the Hamilton Group (Fig. 103). These black shales signal the beginning of a crustal crescendo that re-elevated the Taconic Mountains. The resultant greater relief intensified erosion, sending enormous quantities of detritus (which would form the Hamilton Group). Both terrestrial and marine pebble conglomerates, sandstones, siltstones, and variously colored shales piled up to a thickness of about 11,300 feet. This thick clastic wedge is known as the "Great Catskill Delta." Actually, several deltas became superimposed upon an extensive 1,700 mile coastal plain stretching from Nova Scotia to Georgia.

Fig. 103. Devonian strata in eastern New York.

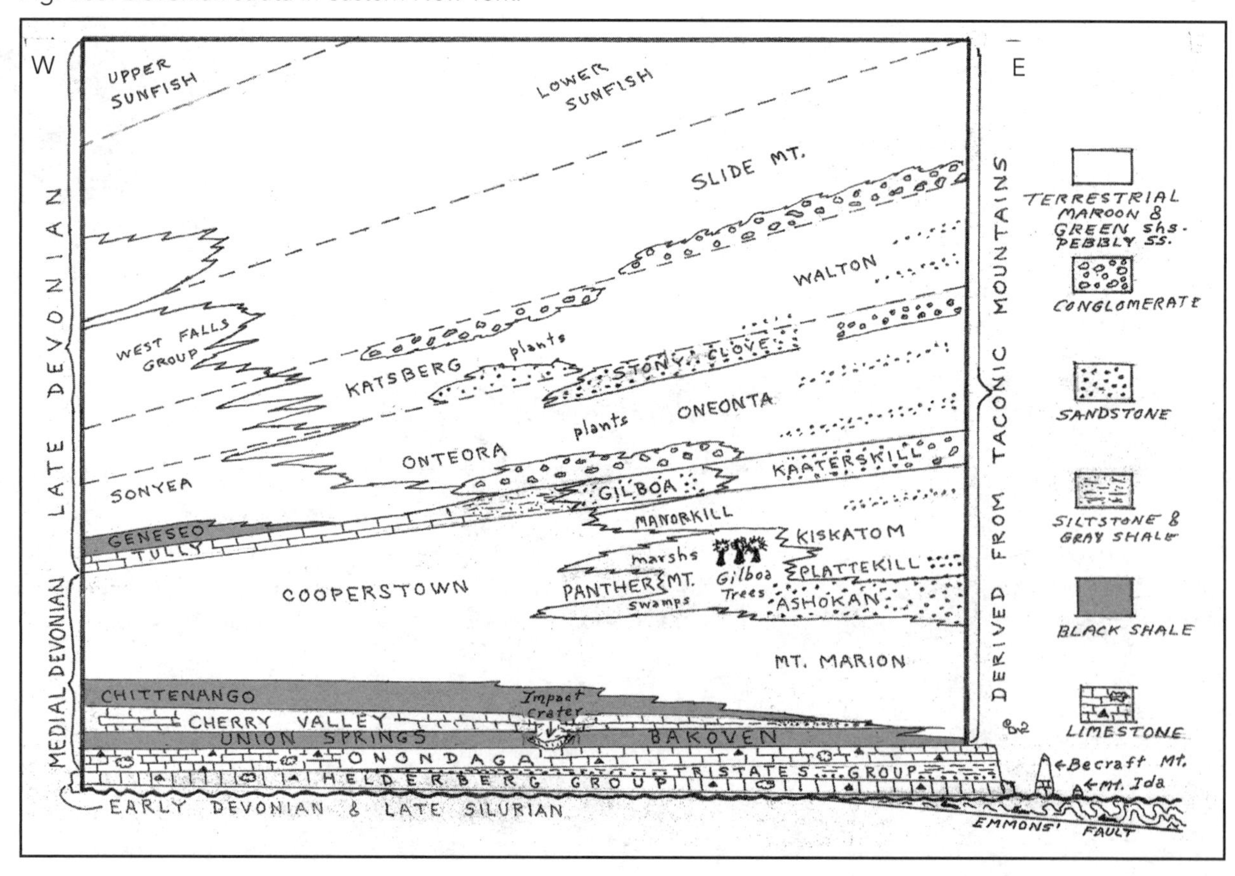

Chapter 24

Late Devonian

382–362 million years ago

In the Finger Lakes region, sandwiched between the subjacent Middle Devonian clastics and the superjacent Upper Devonian clastics, is the Tully Limestone. This carbonate unit represents a brief interlude of crustal stability—the *Tully Tranquility*. The Tully Limestone extends from Chenango County to east of Canandaigua Lake in Ontario County. West of there to near Erie County, pyrite lenses up to 4 inches thick mark the "Tully" horizon. This Leicester Pyrite yields a diminutive Hamilton Group fauna. Never more than 30 feet thick—and usually 5 to 20 feet thick—the Tully Limestone contains horn corals, bryozoans, trilobites, crinoid fragments, and the index brachiopod *Hypothyridina venustula*. Minor components within the limestone are the minerals quartz, tourmaline, muscovite, rutile, and zircon—once again the signature of a metamorphic provenance. Because the Tully Formation presumably grades eastward into the Gilboa Sandstone, up to 325 feet thick, it appears that the Taconics were continuing to furnish some erosional material to the Devonian sea, albeit on a greatly reduced scale. Nevertheless, the position of significant coral-bearing limestone between Middle and Upper Devonian clastics indicates another interim of crustal quiescence associated with a reduction of Taconic-derived material.

During *Phase III of the Acadian Orogeny*, the Taconic Mountains continued their rejuvenation. This residue record is manifest today in the eastern Catskill Mountains, where over a mile-thick accumulation of flat-lying purple and green shales, tan and green siltstones, green crossbedded sandstones, and quartz-pebble conglomerates exist. These are found within the Oneonta, Walton, Slide Mountain, and Sunfish Formations (Fig 103). These Upper Devonian strata constitute the majestic Catskill Front or "Wall of Manitou"—splendidly seen from Cairo to Windham along NY 23—that delineates the western border of the Hudson Valley. Still younger Upper Devonian strata extend westward to Lake Erie. Adding these Upper Devonian Genesee, Sonyea, West Falls, Canadaway, Conneaut, and Conewango groups to the previously mentioned older Devonian clastics amounts to billions of tons of "washed-off" Taconic Mountains, which have been estimated to have reached Alpine elevations of 12,000–14,000 feet.

Additional folding, faulting, foliation, and metamorphism were superimposed upon the older Taconian Orogeny structures, further obscuring the geologic history of the region. Extensive granite intrusions typify the Acadian Orogeny. These intrusions comprise the White Mountains of New Hampshire as well as the major portion of Acadia National Park in Maine. In New York, the Peekskill granite of northern Westchester County may have been injected near the termination of Phase III, perhaps coincident with an extinction event within the very late Devonian.

How did it happen that Devonian seas and lands harbored a multitude of new and advanced forms of plants and animals? Animals can't live without plants (even carnivores require planteaters to survive), but plants *can* live without animals, so there had to be a preceding plant beachhead to start land animal history. Prior to mid-Silurian time, the land was sterile and colored in reds, pinks, purples, oranges, yellows, tans, and browns in the weathered rocks and their decomposition products—soils. But contemporary seascapes were teeming with hordes of organisms. Some of the latter were destined to spill over onto the lifeless land. Salinity and turbidity changes, coupled with overgrazing by the now overcrowded marine animal population of planteaters, left this environment precariously survivable. Land loomed as a haven for both animals and plants. The foremost obstacle to this encroachment was the lack of moist, fertile soils.

In all probability, initial beachheads were tidal marshes, flood plains, and estuaries, where water and nutrient availability could guarantee plant survival. Once established in these transitional environments by adapting to survive in air during low tides and short droughts, necessary structural modifications could be made toward a permanent, progressive occupancy of the land. Over many generations this involved the development of an efficient network of aerial shoots for respiration, and an anchoring root system for water assimilation and structural stability. While immersed in water, plants could be "thin-skinned," for they were within a nutrient-rich medium and needed no special mechanism for respiration—the exchange of the gases oxygen and carbon dioxide; on land it was imperative to develop pores between the cells of the epidermis for this purpose. Ironically, virtually every feature that was an advantage to plants living in water proved to be a disadvantage for survival on land.

The first land plants may have resembled the lichens or the liverworts. A lichen is an intimate intergrowth of two distinct types of organisms—a fungus and an algae. A fungus creates threadlike webs within which grows an alga. While the fungus receives organic food (vitamins) from the alga, the alga reciprocally receives water and dissolved salts from the fungus. Because of this mutually beneficial (*epiphytic*) relationship, lichens were able to extend their habitat beyond an exclusively aquatic life. Nevertheless, their structural limitations remained unchanged, reducing lichens to playing a minor role in the land invasion.

Other plants had greater success in overspreading the land. These included sphenophytes (horsetails and scouring rushes), lycophytes (club mosses), pteriphytes (spore-bearing ferns), and pteridophytes (seed-bearing ferns). These newer plants perfected dense and extensive "pipelines" to furnish mineral-rich groundwater to ever-enlarging branches of leaves. These larger leaf canopies permitted greater light-gathering surfaces for photosynthesis and for eventual resistance and durability to punishing climatic conditions. These advanced plants—yesterday's raw materials for today's coal, petroleum, and natural gas—could only flourish so long as continental areas provided adequate nourishing water, however. Thus, the extent of inland penetration was unavoidably tied to water.

This challenge was met by more extensive development of the seed. Seeds revolutionized fertilization by removing dependence on large amounts of water; seeds could be transported by moving water, wind, insects, and later, birds. Gymnosperms (nude seedbearers) evolved at the beginning of the Devonian Period, but their numbers exploded during the Late Paleozoic and Triassic-Jurassic. Angiosperms (flowering plants) began to fluorish during the Cretaceous Period and Cenozoic Era.

Fig. 104. Medial Devonian plant *Aneurophyton germanicum*, Hamilton Group (Moscow Sandstone).

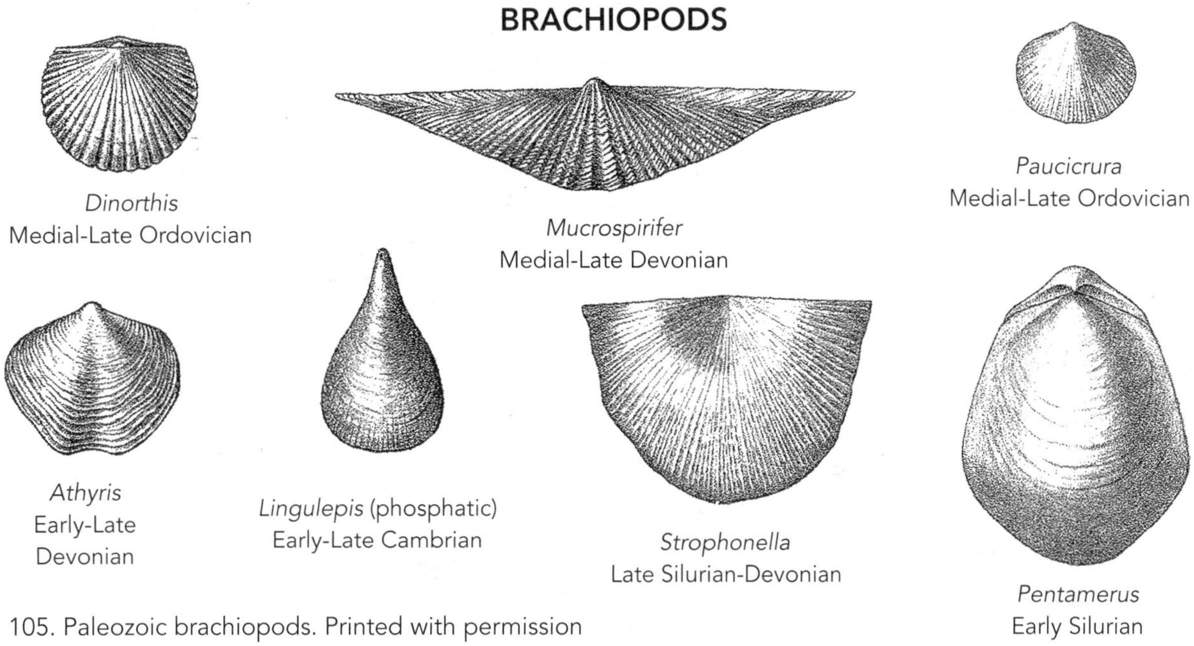

BRACHIOPODS

Dinorthis
Medial-Late Ordovician

Mucrospirifer
Medial-Late Devonian

Paucicrura
Medial-Late Ordovician

Athyris
Early-Late
Devonian

Lingulepis (phosphatic)
Early-Late Cambrian

Strophonella
Late Silurian-Devonian

Pentamerus
Early Silurian

Fig. 105. Paleozoic brachiopods. Printed with permission of the New York State Museum, Albany, NY.

In New York, early land plants include the extinct psilophytes found in the Late Silurian Bertie Dolostone and Medial and Late Devonian psilophytes (*Psilophyton*), lycophytes (*Lepidosigillaria*), sphenophytes (*Hyenia*), pteriphytes (*Tetraxylopteris*), and pteridophytes (*Aneurophyton, Eospermatopteris*) (Fig. 104). The last-named constitutes the earliest known fossil forest, with cypress-like bulbous trunks occuring in three horizons at Gilboa, Schoharie County. These were uncovered by the New York City Board of Water Supply while excavating for the Gilboa Reservoir. (See Hernick, 2003, for additional information.)

Except for the "clicking" of insects, Devonian forests were devoid of animal sounds.

Fig. 106. Medial Devonian starfish (*Devonaster eucharis*), in Mt. Marion sandy shale, along northbound lane of I-87, west of Catskill, Greene County.

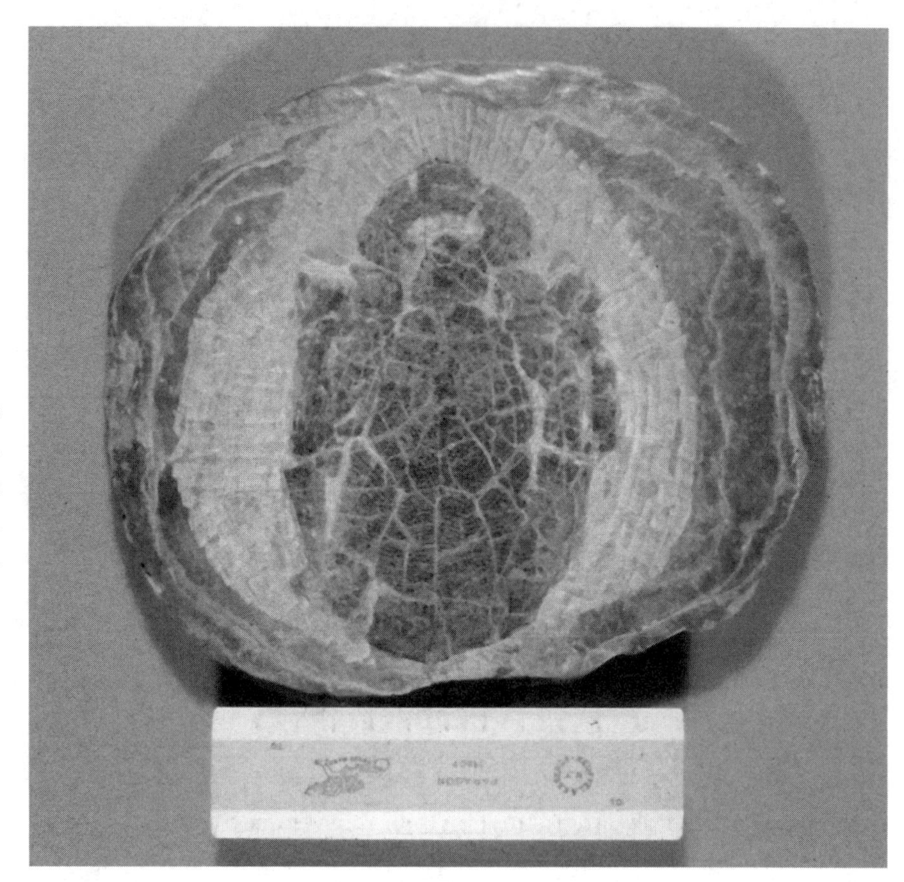

Fig. 107. Placoderm fish from Late Devonian, Scaumenac Bay, Quebec. Printed with permission of the New York State Museum, Albany, NY.

Although the Cambrian Period is noted for the appearance of most invertebrate animal phyla, as far as quantity and diversity within those taxa is concerned, the animal and plant development within the Devonian Period ranks highest within the Paleozoic Era. In the sea, Devonian denizens led both stationary and vagabond existences. Among the fixed inhabitants, the ubiquitous brachiopods (Fig. 105) outnumbered all others on continental shelves. Within this brachiopod[18] shell clutter, athyrids, atrypids, chonetids, orthids, pentamerids, spiriferids, strophomenids, terebratulids, and uncinulids were numerous. Other sedentary shelf-dwellers were bryozoans, corals, sponges, crinoids, and the extinct, crinoid-like cystoids.

In life, however, mobility is an indisputable asset. Moving animals have the choice of picking and choosing their food and selecting a more tolerable environment, and they have the ability to hide or escape from predators. In competing for living space on the sea floor, mollusks that possessed limited mobility (clams, oysters, scallops, and snails) made inroads into the brachiopod domains. In time, burrowers and crawlers became adept at uprooting sedentary occupants. Trilobites—once the varied, perpetual scavengers on Cambrian, Ordovician, and Early Silurian sea floors—now became hapless casualties to the predatory nautiloid- and goniatite-cephalopods, as well as fishes. While myriads of ostracodes flitted from one feeding patch to another, starfishes (Fig. 106) were gorging themselves on oyster and clam beds and sucking the heads of stalked crinoids. Although the eurypterids reached their maximum size during the early Late Devonian—up to six feet in length—their numbers began to diminish. Microscopic, black phosphatic, tooth-like conodonts became so abundant that they, with the cephalopods, form the biostratigraphic zonation for the Devonian Period.

Another major category of marine life was fishes. Five classes of these early vertebrates spawned in Devonian seas and lakes: jawless fishes (ostracoderms); spiny fishes (acanthodians); armor-plated fishes (placoderms); cartilaginous fishes (sharks, rays); and bony fishes, including two groups of lungfishes (Dipnoi) and lobe-fins (Crossopterygii). Fish became so varied and abundant that it is appropriate to dub the Devonian the "Age of Fishes."

Ostracoderms, the most primitive of fishes, had well-developed bony head armor and a scale-covered flat body. Lacking paired fins, they probably wiggled tadpole fashion, and without jaws—like modern lampreys—they rode "piggyback" by affixing their suction-cup mouths to swimming shelled cephalopods or other fishes. The evolution of jaws in the remaining four classes enabled these fishes to expand their arena of activity, vary their diets, and fend off attackers.

Probably the scaled and spiny acanthodians were the first fishes to use jaws effectively, as did their companions, the placoderms (Fig. 107), who developed thick skin plates of armor. As the placoderms acquired more offensive equipment and the ability to move more rapidly, however, the need for defensive armor diminished. Like the knights of old, placoderms eventually shed their armor for "chain mail" (overlapping scales), evolving into the bony fishes, which are so abundant in our marine and nonmarine waters today. Thus, even in Devonian times, the best offense was a good defense.

Cartilaginous fishes (sharks and rays) are prime examples of "permanent identities"—modern sharks and rays appear to be nearly identical to Devonian ones. They are characterized by an almost complete absence of bone, and an internal cartilaginous skeleton. They also possess perfectly aquadynamical bodies optimized for speed and maneuverability. Combined with their powerful jaws studded with razor-sharp teeth, these animals ruled supreme in their unchallenged kingdom. There was, and remains, little impetus for evolutionary change.

Among the bony fishes with overlapping scales and internal skeletons of calcium carbonate, two groups deserve particular mention. The Dipnoi ("double-breathers") utilized respiration by gills when submerged in water and by lungs when buried in mud during drought—thereby demonstrating their capability to survive in both realms. The Crossopterygians ("lobe-fins")—an evolutionarily important fish—featured paired limbs that were short, muscular, and with bone configurations akin to the limbs of amphibians, reptiles, birds, and mammals. During the Devonian, two divisions of crossopterygians evolved. One, the rhipidistids (*Eusthenopteron*), crept out of the ocean to become amphibians—one notable example being the Late Devonian *Ichthyostega*. This, and the Late Paleozoic eight-foot monster salamanders, had large eyes, splayed-out legs and long slithering tails. Their fossil dental records indicate a diet of insects and beached fish—not plants. The other crossopterygians were the coelacanths. They remained in saltwater. Today, *Latimeria* still survives in the western Indian Ocean near the Comoro Islands, virtually unchanged from its Devonian ancestor.

We can imagine that one Devonian day when an adventurous rhipidistid "walking fish"—either fleeing from a predatory shark, or ogling an insect morsel on a partly submerged log, or washed ashore during a storm—unexpectedly found itself on land, and survived. It was the pioneer of a vanguard that would establish a foothold leading to a mass animal invasion of the land.

A team of paleontologists on Ellesmere Island in arctic Canada recently discovered an evolutionary link between fish and amphibians, named *Tiktaalik roseae*. The specimens, from 4 to 9 feet long, have the limbs, skull, neck, and ribs of four-limbed animals, but also the primitive skull, neck, fin, and scales of fish. The fossils are 375 million years old in Medial Devonian strata.

Summary of Acadian Orogenesis

1. Like the Taconian Orogenesis (Phases I, II, and III), the Acadian Orogenesis (Phases I, II, and III) began with *tranquilities*—the Helderberg, Onondaga, and Tully—except that the Helderberg carbonate deposition occurred over longer duration than either the Onondaga or Tully carbonate deposition.
2. Phase I detritus was relatively thin as compared to that of Phases II and III.

Chapter 25

A Devonian-Age Impact Crater

Buried beneath the Catskill Mountains is evidence of an ancient collision with an extraterrestrial body (Fig. 108). This oddity is believed to be a 7-mile-wide meteorite impact crater now buried under Panther Mountain in northern Ulster County, west-southwest of Phoenicia.

In 1950, George H. Chadwick, an avid researcher of New York Devonian geology, noted the anomalous circular course of Esopus Creek relative to Panther Mountain. He referred to this as a "great rosette," interpreting it as a low dome. This structural interpretation piqued the curiosity of oil and gas geologists, and suggested a domal trap for possible accumulation of petroleum and/or natural gas. In 1955 the Herdman exploratory well was drilled. At a depth of 4,350 feet, about 50,000 cubic feet of gas was found (Kreidler, 1959). But after a few days, production dropped dramatically and the deepened hole was abandoned at a depth of 6,400 feet, where it had penetrated the Middle-Upper Ordovician Schenectady Formation.

Almost three decades later Isachsen, Wright, Revetta, and Dineen (1977) suggested that the circular drainage pattern peripheral to Panther Mountain might be the rim trace of a buried meteorite crater. Additional research by Isachsen, Wright, and Revetta (1994) has furnished evidence that Panther Mountain is indeed sitting on a hidden crater created by a meteorite or comet impact. Their evidence includes:

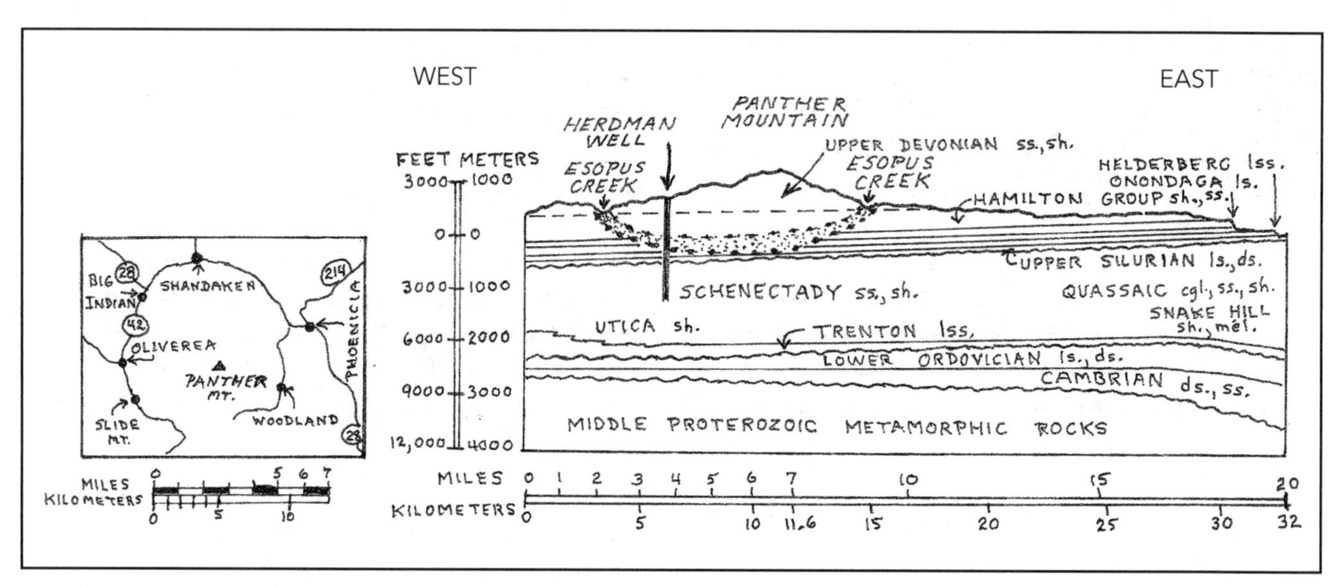

Fig. 108. Strata of the Panther Mountain region of the Catskill Mountains.

- *An overabundance of bedrock fractures (joints)—more than ten times the normal amount—is found in the flat-lying strata of Esopus Creek.* Whereas the preponderance of joints rimming Panther Mountain is not proof of meteoritic impact, it is evidence of severe bedrock breakup in an area outside of brittle deformation caused by the Acadian Orogeny.
- *Lower density rock, as measured by a gravitometer, exists directly under Panther Mountain.* This indicates greater porosity rock than that of the surrounding identical rock formations. Fragmentation of the rock from the explosive energy released during the impact could account for the increased porosity.
- *Drill samples from the Herdman well contain pinhead-sized quartz spherules.* Owing to the tremendous energy released upon impact, a meteorite, along with the rock it strikes, will liquify and vaporize. In this case, melting of the bedrock would produce a rain of silica (quartz) spherules that would settle in the vicinity of the impact site.
- *Presence of silica shock quartz shards and quartz lamellae.* A meteorite impact would induce a sudden devastating shock on subsurface strata, causing them to shatter into incoherent patterns. The resulting material is known as "shock quartz." In addition, traces of melted and unmelted quartz appear as parallel lines—called lamellae—along microscopic planes within the quartz.

All of the above clearly point to an impact event. Some unanswered questions remain, however:

1. If a meteorite crater lies buried beneath Panther Mountain, was it caused by a stony meteorite, a metallic meteorite, or a comet?
2. What is the depth of the impact crater below the covering strata?
3. When did the impactor strike the Earth?

In response to question #1, no meteorites have been reported. This is not surprising because impacts result in almost complete vaporization, and any surviving meteorites would be deeply buried. As to question #2, the estimated depth of the crater from the summit of Panther Mountain (elevation 3,760 feet) is approximately 3,800 feet.

Over 50 years ago the author discovered a rock exposure that today has relevance to question #3—

Devonian Days Were Shorter

An interesting bit of paleontological research serves as an example of the intertwining of sciences. During the 1960s, paleontologist Dr. John W. Wells of Cornell University discovered that fine growth lines on coral exoskeletons represented daily growth increments; i.e., the coral secreted one thin layer of calcium carbonate each day. Coarser monthly bands and still thicker yearly bands were also observable. Using several species of living corals, Wells counted an average of 360 lines per year, close to our 365 days per year. Transferring his attention to Medial Devonian corals, Wells found an average of 398 daily growth lines per year, which indicated that Medial Devonian years numbered about 400 days. Wells' trend curves suggested Earth's rotation slowdown to be 1 day for every 10.8 million years. Extrapolation of his data reveals that an Early Cambrian year contained 416 days and a Medial Ordovician year had 408 days. The dinosaur *Triceratops* experienced a Late Cretaceous year of about 372 days.

Astronomers have long theorized that the tidal action of our moon has continually slowed the Earth's rotation. Ironically, the work of astronomers—accustomed to looking "upward" to the skies—has been solidly reinforced by geologists and paleontologists who look "downward" to the Earth.

Fig. 109. Close-up of pseudo-cleavage zone in upper Union Springs Shale, road cut along south side of Otsego County 54, 2.7 miles east-northeast of Cherry Valley.

the time of impact. While mapping the bedrock geology of the Mohawk Valley, a fresh outcropping (along what is now Otsego County 54) was excavated during the relocation of US 20, 2.7 miles east-northeast of Cherry Valley. The rock displayed a peculiar appearance—between horizontal strata of the Union Springs black shale was a disturbed zone (approximately two feet thick) of convoluted shale and limestone concretions with various attitudes. Beneath this was a shale interval (approximately one foot thick) containing closely spaced, nearly vertical fractures resembling cleavage (Fig. 109). Such cleavage denotes great pressure. This was similar to regional cleavage so prominent in older shales observed in the Taconic region (Fisher, 1980). At that time, interpretation was perplexing; the author suggested a violent sediment disturbance created by submarine earthquake activity. Now, in light of the Panther Mountain findings, an alternate interpretation is plausible. The convoluted shale, disoriented concretions, and brittle fracture zones that the author observed in the Union Springs Shale were created during the assumed Panther Mountain impact event. If correct, the age of the hypothetical subsurface Panther Mountain crater is Medial Devonian. More specifically, the impact was contemporaneous with deposition of the Union Springs Shale Member of the Marcellus Formation of the Hamilton Group. This would date the impact as occurring about 390 million years ago; accordingly, covering strata of the crater would be the remainder of the Hamilton Group (upper Marcellus, Skaneateles, Ludlowville, and Moscow Formations). Estimates of up to 10,000 feet of Middle and Upper Devonian clastics above the crater have been postulated.

Chapter 26

Erasure by Erosion, and the Alleghenyan Orogeny

The Mississippian–Pennsylvanian–Permian Trilogy

362–245 million years ago

How does one restore the geologic history of a time interval that is virtually unrepresented by a rock record? That gap is a lengthy one in New York State; there is an almost complete rock-record vacuum for the last 117 million years of the Paleozoic Era. Only sporadic exposures of a quartet of thin Mississippian and Pennsylvanian-age strata occur in Allegheny, Cattaraugus, and Chautauqua counties in southwesternmost New York. Because of this deficiency, we must look to neighboring Pennsylvania and West Virginia, where thick Upper Paleozoic strata furnish a plentiful database for reconstructing these periods in New York State.

In Pennsylvania, Mississippian-age sandstones are found in the Pocono Mountains, and Pennsylvanian-age black shales and coal beds yield luxuriant and diverse fossils of ferns, horsetails, and club-moss trees. These are all indicators of extensive swamp-jungle environments, providing clear evidence that these localities were at tropical latitudes. Today, this flora grows on the periphery of plant ecosystems. The modern, swampy Florida Everglades provides a dwarf analogy to the Late Paleozoic mega-swamps of Pennsylvania, West Virginia, and Illinois. Those ancient, dank, and steamy swamps were home to bulky, 8-foot amphibians, giant dragonflies with wingspans of two feet, foot-long cockroaches, and oversized centipedes and millipedes.

The end of the Permian Period was marked by the most dramatic and thorough extinction in the Earth's history. Recent evidence suggests that a tremendous volcanic eruption occurred in what is now western Siberia, resulting in a "one-two punch" of global refrigeration (caused by increased reflection of sunlight) followed by dramatic and rapid greenhouse heating (caused by increased CO_2 released in the eruptions). This environmental disaster was rapid and extreme—and few species survived. But those that did, flourished, in part because of evolutionary advancements of the egg—insuring perpetuation of their lineages during water and food famine on land.

Considering the meager Late Paleozoic rock record in New York State, is there any evidence that the aforementioned flora and fauna lived here? Judging from the existing elevations and dips of Mississippian and Pennsylvanian-age strata in neighboring Pennsylvania and Ohio, it is reasonable to suppose that these fossil-rich rocks formerly extended well into southern New York. Simonsen and Friedman (1991) suggest that these overlying rocks had a thickness of 1.7–3.5 miles atop the present Devonian rock section. Likewise, Ervilus and Friedman (1991) reported that subsequent to the deposition of the Middle–Upper Cambrian Pine Plains Dolostone, southeastern New York was buried to a depth of 3.1 miles. Obviously, specific ages of this now-removed overburden are unknown. Relicts of proven plant fossil-bearing Pennsylvanian–Permian strata occur, however, in the Naragansett Basin of Rhode Island and the Worcester, Massachusetts, coal deposit, and were also uncovered during a building excavation at Pelham Bay, New York. These statements support the idea that a vast volume of Upper Paleozoic rock has been stripped away by erosion. The prolonged and elevating *Alleghenyan Orogeny* implemented this extensive obliteration of sedimentary rock. Extending from Nova Scotia to Alabama, this intense fold and reverse-fault mountain-building episode finalized the construction of the Appalachian Mountain chain.

During the Late Paleozoic Era, were the Taconic Mountains jutting up through a sedimentary blanket of hundreds or thousands of feet of strata, or were they buried beneath it? In the southern Hudson River Valley, any alleged Alleghenyan overprinting of Acadian and Taconian structures is not readily apparent and not easily proven. Lacking conclusive evidence of folding and reverse-faulting of Late Devonian strata, structures arising from the Alleghenyan Orogeny would escape detection. Nevertheless, there are some geologic clues in this region that hint toward probable Alleghenyan folding. Perhaps most convincing is the noticeably northeast-southwest structural trend of the Shawangunk Mountains and the parallel-trending Middle Proterozoic metamorphic rocks of the Hudson Highlands. The trends of both upland regions are counter to the more northerly-striking Acadian or Taconian structures observed within the Taconic Mountains.

In the Greenwood Lake Quadrangle in Orange County is a similar northeast–southwest trending tight syncline of Oriskany Sandstone, Bellvale Sandstone, and Schunnemunk Conglomerate—all alleged to be Medial Devonian in age. This syncline lies in a rift flanked by presumed Late Triassic–Early Jurassic *high-angle normal faults*. Owing to the differing fold alignment, a tentative Alleghenyan fold to the Greenwood Lake syncline is plausible.

The age and magnitude of crustal rearrangement attributed to the Alleghenyan Orogeny may be estimated from radiometric dates obtained from a belt of igneous plutons—chiefly granites found in eastern Massachusetts, New Hampshire, Maine, and Montreal, Quebec. Plutonic rocks are not formed at the Earth's surface; rather, they form at depths of over 20 miles, where pressures and temperatures are intensely high. Therefore, their current elevations above sea level indicate extensive uplift. These granite intrusions, dated as Pennsylvanian and Permian ages, provide additional evidence of Late Paleozoic structural rejuvenation. Consequently, the case for a Late Paleozoic sedimentary blanket concealing the Taconic Mountains, followed by a nearly maximum erosional erasure, deserves serious consideration.

From the close of the Paleozoic Era, at least another 150 million years would elapse before erosion would mark the initial incising of the Catskill Plateau.

Tectonic Plates Disperse

Chapter 27

A New Face for an Old: 245 Million Years of "Aging"

The Uplifting Mesozoic and Cenozoic Eras

Unlike the previous Paleozoic Era in eastern North America (which was characterized by multiple crustal *compression* events with associated orogenies), the succeeding Mesozoic Era (Triassic, Jurassic, and Cretaceous Periods) witnessed the opposite—crustal *tension*, or "pulling apart." A geologic tensional event is termed a *taphrogeny* and is characterized by subsidence, rift phenomena, and high-angle normal faulting. Pangea, the supercontinent that assembled during the Permian Period, began to split at the beginning of the Mesozoic. This initial breakup would eventually create the continents we see today and a new ocean—the incipient Atlantic. This taphrogeny created a Triassic–Jurassic faulted landscape of down-dropped lowlands (grabens) and basins (rifts) alternating with uplands (horsts). This "up and down" topography was crisscrossed with a network of high-angle normal faults, usually displaying great vertical displacements. In easternmost North America, this tensional episode is termed the *Palisadian Taphrogeny*, named for the stockade-like Palisades Sill (a vertically jointed columnar volcanic intrusion) overlooking the Hudson River in Rockland County, New York (Fig. 110), and extending into New Jersey. Using radiometric dating, it has been determined that the Palisades Sill was emplaced 195 ± 5 million years ago (Erickson and Kulp, 1961)—near the boundary of the Triassic and Jurassic Periods.

Like the late Paleozoic, the Mesozoic is poorly represented in the rocks of New York State. You will not find the bones of *Tyrannosaurus rex* in New York strata! For a more complete documentation of this fascinating "Age of the Dinosaurs," we must look to the western United States.

From Nova Scotia to North Carolina, Late Triassic–Early Jurassic rift valleys were clogged with thick, terrestrial sediments interlayered with blankets of basalt flows. Some valleys were floored with freshwater lakes in which black muds accumulated; today the resultant shales contain the fossils of fish and reptiles. In Massachusetts, Connecticut, New York, New Jersey, and Pennsylvania, the nonmarine Newark Group (Stockton, Lockatong, Brunswick, and Hammer Creek Formations) is composed of arkose (a feldspar-rich sandstone), sandstones, siltstones, and shales variously colored in shades of maroon, purple, pink, and orange (Fig 110). In New York, New Jersey, and the Connecticut River Valley, extensive rock and mudslides draped precipitous fault-formed cliffs. These alluvial fans and talus deposits partly concealed the fault scarps. In Rockland County, New York, the northeast-trending, intermittently active Ramapo Fault delimits the Newark Group on the northeast abutting the Hudson Highlands gneisses. A Palisadian normal fault is believed to form the east margin of Stissing Mountain in northern Dutchess County (Fig. 111). It is also conceivable that the Mohawk Valley normal faults, which created a horst and graben topography, may have originated or been reactivated during the Palisadian Taphrogeny. None of these faults cut uppermost Silurian strata, but they disappear to the south within the thick Middle–Upper Ordovician Utica and Schenectady formations. This implies an Early Silurian tensional event following

Phase V of the Taconian Orogeny. Undoubtedly, further study will reveal more Palisadian faults in the northeastern United States.

Naturally enough, Mesozoic fossils in New York are as poorly represented as the rocks. One dinosaur that is known to have lived in New York was *Coelophysis* (see-low-fy'-sis)—a bipedal, 9-foot-long, 85-pound carnivore (Fig. 112). In 1972 high school students Paul Olsen and Robert Salvia discovered footprints of these animals at Nyack, Rockland County (Fig. 8) during the excavation of a building. Identification of the footprints[19] was made by Dr. Edwin Colbert, formerly of the American Museum of Natural History in New York City. The raptor *Coelophysis* patrolled riverbanks and lake shores seeking prey concealed within scattered clumps of ferns, horsetails, cycads, and primitive pines. It is assumed that other dinosaur species inhabited New York; however, to date no fossil evidence has been found. See Fisher (1981) for a more comprehensive discussion of the flora, fauna, and geology of southeastern New York 200 million years ago.

The only other Mesozoic strata in New York are unconsolidated or semiconsolidated Late Cretaceous clays and sands on Staten Island, and limited exposures on the north shore of Long Island. During the Late Jurassic and Cretaceous periods, an elevated tableland—the Allegheny Plateau (from the Taconic foothills westward to the West Virginia area)—formed as a result of regional uplift (*epeirogeny*). The Canadian Shield

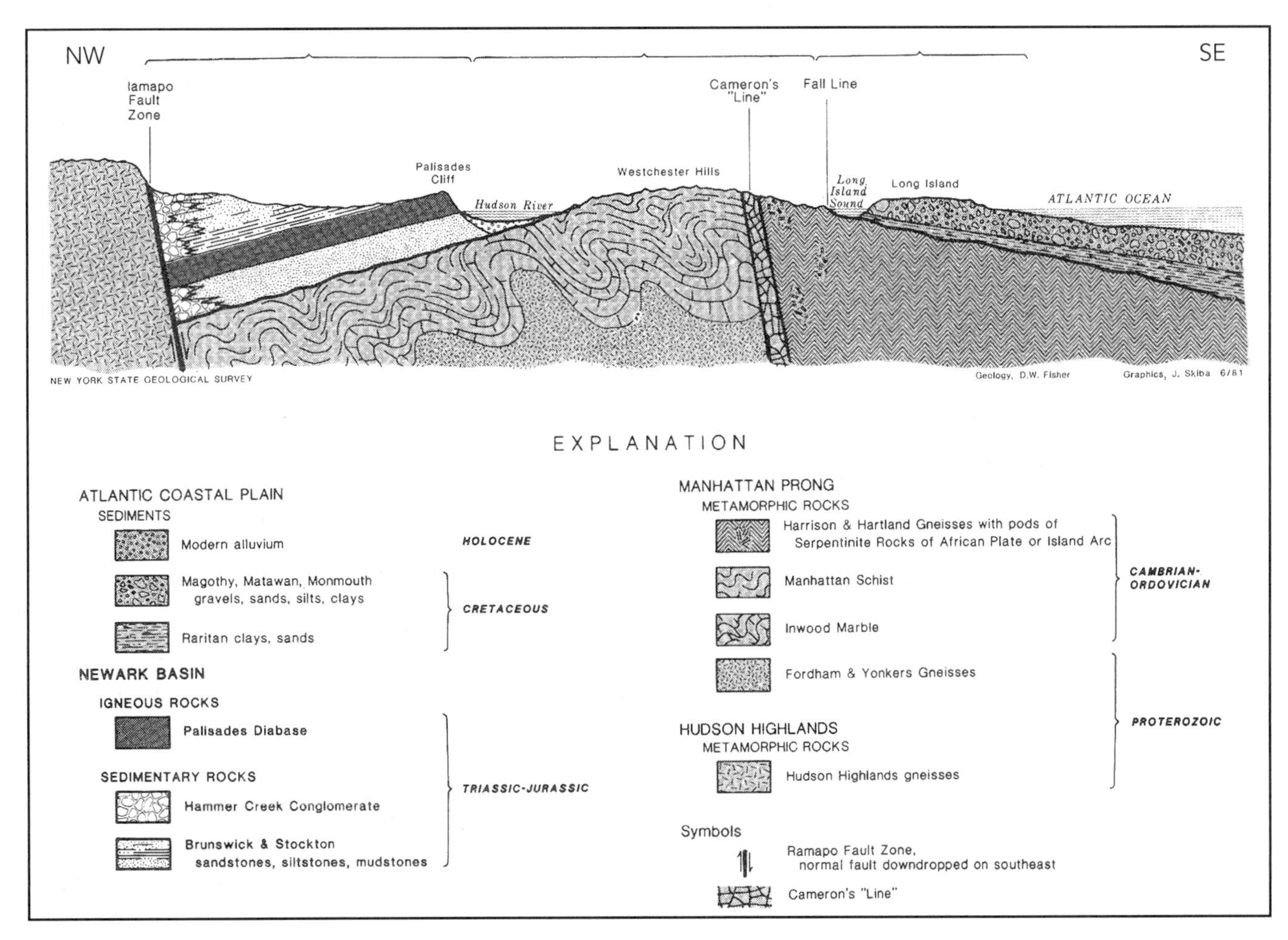

Fig. 110. Generalized structure profile across the Newark Basin, from the Hudson Highlands to the Atlantic Ocean. Cartography by John B. Skiba. Printed with permission of the New York State Museum, Albany, NY.

and the site of the future Adirondack Mountains were probably also uplifted during this event. The effect of this epeirogeny in the Taconics was not uplift as a unit; rather, a series of semiparallel hogbacks resulted. A dendritic pattern of headward erosion by streams and rivers dissected the "layer-cake" Devonian and presumed Upper Paleozoic strata and began sculpturing for the forthcoming Catskill Mountains. The ancestral Hudson River probably came into existence during the Cretaceous Period. The courses and proliferation of rivers and streams were a meshwork of vertical fractures intersecting at various angles. Near Peekskill in Westchester County, the course of the Hudson River has a "doglegged" appearance caused by two sets of joints intersecting at approximately 30°. Erosional detritus from the elevated Allegheny Plateau formed an extensive apron along the southeastern front of the Appalachian Mountains. This clastic wedge comprises

Fig. 112. Late Triassic hollow-boned raptor *Coelophysis*; 8 feet tall, 85 pounds. Drawing by Kathryn M. Conway, based on trackways found near Nyack, Rockland County, and skeletons from the Chinle Formation, New Mexico. Printed with permission of the New York State Museum, Albany, NY.

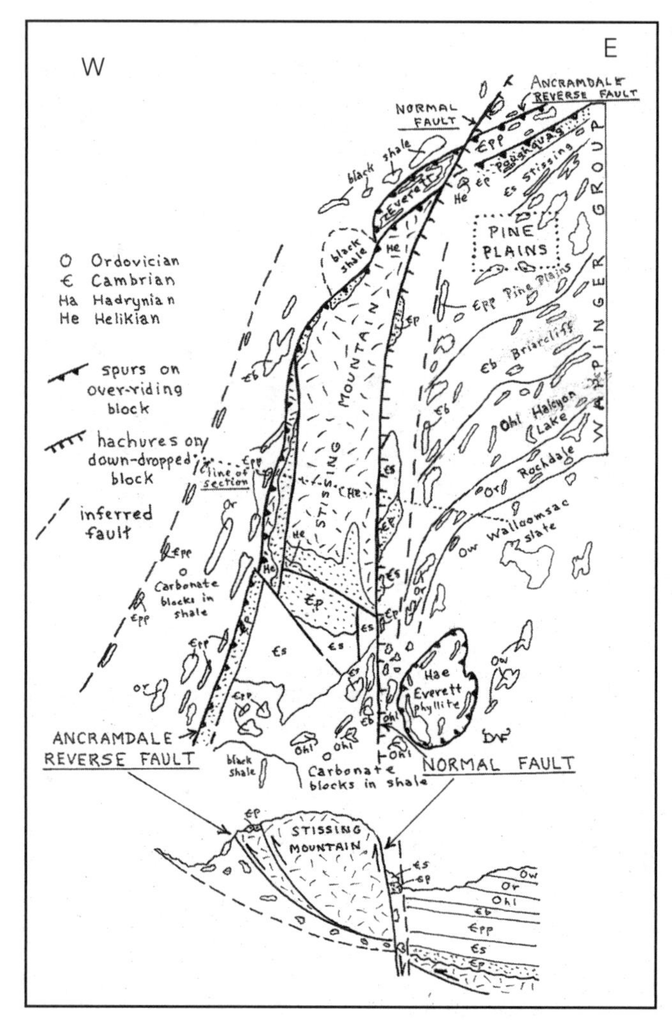

Fig. 111. Bedrock geology near Pine Plains, Dutchess County.

the Atlantic Coastal Plain with its northern boundary at the "Fall Line"—the physiographic juncture of higher-elevation Paleozoic bedrock to the west, and lower Mesozoic-Cenozoic sediments to the east.

Pre-Pleistocene Cenozoic Era sedimentary and paleontological evidence is unavailable in New York State. If any Eocene through Pliocene deposits were laid down, they were removed by the subsequent bulldozing of Pleistocene glaciation. In all probability, widespread regional uplift continued after the Cretaceous Period. Such increases in elevation fostered recurring erosion, principally by increasing stream gradient. This erosion imprinted New York's basic drainage patterns on the exposed Allegheny Plateau; the eastern portion of this deeply "dissected" plateau is observed today as the rugged topography of the Catskill Mountains.

Chapter 28

A Cracked Dome: the Adirondack Mountains

New Mountains from Old Rocks

Neogene Period: 25 million years ago–today

Simultaneous to the continuing development of the Catskill Mountains topography, another mountain range was being "born" in northern New York—the Adirondacks. This mountain range is a 125-mile-diameter dome and an isthmus-like extension of the Archean and Lower–Middle Proterozoic Canadian Shield, connected to that shield in the Thousand Islands region of the St. Lawrence River (Fig. 4). The Adirondacks are composed of some of the oldest rocks in New York State—1.34 billion-year-old plutonic metanorthosite, metagabbro, charnockite, and 950 million-year-old granite. These Middle Proterozoic igneous and metamorphic rocks are estimated to have formed 25–30 miles beneath the surface.

How did these deeply buried rocks reach their present lofty elevations? The mechanism that caused this uplift is speculative. One possibility is that a mantle "hot spot" exists under this region. A "hot spot" is a thinning of the Earth's crust, which allows the higher-temperature mantle to more closely approach the surface. The high temperatures associated with the hot spot can cause expansion and uplift of the overlying crust. Upward movement of these once deeply buried "ancestral Adirondack" rocks created an acceleration of weathering and removal of the once-overlying Paleozoic sedimentary blanket.

Evidence indicating that Paleozoic rocks once covered the Adirondacks can be seen along NY 30 at Wells in southern Hamilton County. Here, fossil-bearing strata of Late Cambrian through Medial Ordovician

Fig. 113. Schematic cross section of the Adirondack region.

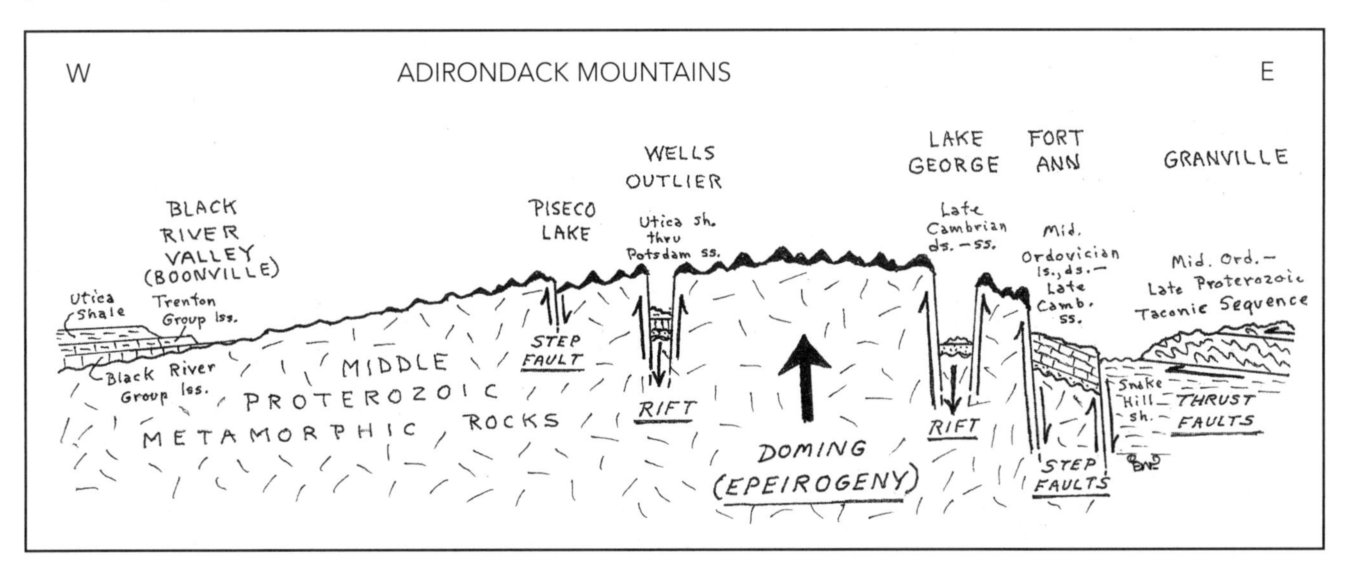

ages are exposed in a rift well within the periphery of the present Adirondack Mountains (Fisher, 1957). Geologists refer to these isolated remnants of once-continuous strata as "outliers"—the unit described here is the "Wells Outlier" (Fig. 113). Additional Upper Cambrian through Middle Ordovician remnants are exposed in the rift valley occupied by Lake George on the Warren–Washington County line. While no exposures of Silurian or Devonian age have been identified in any outliers, projections of these strata located south of the Mohawk Valley suggest at least partial Adirondack area coverage during these times, with subsequent removal by erosion (Fig. 113). (See Silurian and Devonian maps.)

Recent elevation measurements of benchmarks (permanent markers placed on selected geographical points such as mountain peaks) indicate that the Adirondacks are currently rising at a geologically "breakneck" speed of ⅛ inch per year. This continuing uplift results in frequent, but minor, earthquake activity, especially in the central Adirondack Mountains near Blue Mountain Lake in Hamilton County.

Based on estimates of the depths at which the Adirondack rocks were created, it would seem relatively easy to calculate how long they have been rising—i.e., when the mountain range began to form. Unfortunately, however, there is no way of knowing what the rate of uplift has been in the past. An estimate of the onset of Adirondack doming (a term used to describe the uplift of a large, roughly circular area—also referred to as *epeirogeny*) is therefore only a rough approximation. We can reliably assert that the forces pushing the Adirondack rocks upward were in effect during the Miocene Epoch (25–10 mya), and these mountains reached a maximum elevation during the Pliocene Epoch (10–2 mya). Within the subsequent Pleistocene Epoch, however, continental glaciation (to be discussed more completely in the following chapter) abraded and beveled the jagged mountains into a smoother and diminished topography—much as we see today.

While the doming (*epeirogeny*) was a critical event in the forming of the Adirondacks, another force—*tension*—was equally important in shaping the topography of this range. Tension is the "pulling apart" of rock, and frequently results in normal faulting; a geologic action that combines tension with normal faulting is referred to as a *taphrogeny*. Evidence indicates that during the creation of the Adirondacks, epeirogeny and taphrogeny were occuring simultaneously.

Within the eastern and southern Adirondacks, numerous faults have fractured Middle Proterozoic metamorphic rocks and Upper Proterozoic basalt dikes. In the adjacent Champlain and Mohawk valleys, normal faults (possibly active during Triassic age) have cut Cambrian and Ordovician sedimentary rocks. Major faults within the Adirondack dome increase in vertical displacement to the north and northeast, indicating that a large scale "hinging" of the dome movement occurred to the south and southwest.

Mafic and ultra-mafic dikes found in the Champlain Valley are of particular importance in determining when normal faulting occurred in the Adirondack region. Radiometric analysis of these dikes indicates that these intrusives formed 116–106 mya, based on rubidium-strontium ratios. Other isotopes (potassium-argon) yield dates of 114 mya. Since faulting of a dike can only occur after intrusion takes place, the faulting of some of the dikes occurred after Early Cretaceous time.

Deciphering the geologic evidence in order to comprehend the events and chronology of the formation of the Adirondacks is a frustratingly complex challenge. To summarize our current understanding: initial normal faulting in the Adirondack region was established during the Late Proterozoic Era and reactivated during the Triassic and Early Cretaceous periods. Fracturing attained a culmination during the doming in the Miocene and Pliocene epochs of the Neogene Period. Doming continues today, and persistent earthquake activity in the region provides evidence that faulting may still be occuring.

Interestingly, the brunt of this "Adirondackian" upheaval coincides with the early stages of mountain building on the opposite side of the planet, where the Indian Plate continues to collide with the Eurasian Plate. This has created the highest mountain range—the Himalayas—and the highest plateau—the Tibetan Plateau—found on Earth today.

Having established that the uplift of the Adirondack region is geologically quite recent, and that the rocks comprising the region are among the state's most ancient, we conclude that New York's *youngest mountains* are composed of New York's *oldest rocks*!

Chapter 29

A Lengthy Winter and the Modern Look

The Pleistocene Epoch: 2 million years ago–today

"How did rock fragments and boulders of northern origin get to their present positions? ... that these phenomena would be explained by the assumption that polar ice once reached to the southernmost edge of the district now covered by these rock remnants ... and in the course of thousands of years gradually melted back to its present extent." (Reinhard Bernardi, first scientist to propose the idea of a glacial period in the Earth's history, 1832.)

Unlike the Paleozoic tectonic events, which involved varying degrees of heat, the most recent revolutionary episode that remolded the Taconic, Adirondack, and Catskill mountains and most of New York State was a "big chill."

Swiss naturalist Louis J. Agassiz (1807–1873) is traditionally credited as the first to recognize the former existence of an ice age; however, his initial publication proposing a glacial epoch (*Études sur les Glaciers*, 1840) post-dates the report of Bernardi. Agassiz's thesis was based on his own observations and on those of other geologists. These men reported on the existence of polished bedrock, *exotic* boulders ("out-of-place" boulders that often don't match surrounding bedrock; also known as glacial *erratics*), and groupings of parallel grooves or *striations* in the bedrock (Fig. 114). Agassiz proposed that "the earth was covered with an immense sheet of ice, in which the mammoths of Siberia were buried and which extended to the south as far as the phenomena of erratic blocks ... filling the Baltic Sea and all the lakes of Germany and Switzerland ... and covering even all of North America." Later, it would be demonstrated that not all of North America was ice-covered—only the northern half.

Fig. 114. Glacial striations and polish on Hamilton Group sandstone, near Catskill, Greene County. Photograph by Steve Nightingale.

Agassiz came to the United States in 1846; the following year he was appointed professor at Harvard University, where he founded the famous Museum of Comparative Zoology. Although his major contributions to science were in the realm of fossil fish (Paleoichthyology), Agassiz's principal fame lies with his revelation of clues to a past ice age.

Because polar regions regulate much of the Earth's weather, ice-covered regions are a force to be reckoned with. Currently, intensive research in Alaska, Antarctica, and Greenland is permitting better understanding of the mechanisms and results of past ice ages. Valid evidence, including glacial deposits and striations on bedrock, reveal that ice ages have periodically taken place for the last 2.5 billion years. Exactly when the most recent glaciation began is controversial, but we do have ice core evidence of ice caps existing in Antarctica 38 mya. Was this the beginning of the ice age that affected the world during the Pleistocene Epoch, when it has been estimated that 40 million km^3 of ice accumulated on almost one-third of the Earth?

Beginning during the Mid-Cenozoic Era (Oligocene Epoch), world climates steadily grew colder, with average temperatures reaching a minimum during the Pleistocene Epoch. Among several source areas for advancing ice sheets, northern Quebec east of Hudson Bay gave rise to the ice that resculptured the northeastern United States. This ice pileup crept in all directions, alternately advancing and receding several times as a result of cold glacial and warmer interglacial intervals. (It is important to realize that glacial ice cannot change the direction of its flow. There are two factors that affect the direction of the ice *front:* accumulation of snow pack on the ice tends to push the ice forward, while ablation [melting] along the ice front causes the leading edge of the ice to recede. If the mechanisms are in balance, the ice flows forward while the leading edge remains stationary.)

Some research indicates that there may have been as many as 30 severe glacial periods in North America, but until a few years ago, it was widely accepted that North America experienced four major glacial advances, of which only the last (termed the Wisconsinan Stage) was recognized as having affected New York State. Here, the ice blanket probably reached a thickness of over 5,000 feet, except for an unglaciated area in Allegheny State Park in Cattaraugus County. Also noteworthy is the fact that Long Island did not exist at this time; that locale was a continental shelf floored with Late Cretaceous and Early Cenozoic sediments, and under as much as 300 feet of ocean.

There are a number of possible reasons for global cooling. One of the most obvious is the relocation of tectonic plates from tropical or temperate zones to polar zones. Geologists have unmistakably demonstrated that throughout Earth's history tectonic plates have occupied differing latitudes. For example, when Taconic sediments were deposited, North America was positioned in the Southern Hemisphere between the equator and the Tropic of Capricorn. During the Mesozoic Era, Antarctica was not at the South Pole—it was in tropical or semi-tropical latitudes. Evidence for this includes abundant fossils of land plants, coal beds, and bones of warm-weather dwellers. However, even though "icing over" is an expected result for a plate moving into polar regions, it cannot completely account for the Pleistocene Ice Age. The movement of the North American Plate is too slow to accommodate the geologically "quick" occurrence of Pleistocene glaciation.

In the nineteenth century it was proposed that the energy output from the sun might vary to create cycles of higher and lower average temperatures. Current advances in scientists' understanding of nuclear fusion in the sun effectively rule out this possibility. A more plausible cause could be the presence of a persistent, globe-encircling shroud of volcanic ash and dust that would screen out solar energy and reduce global temperatures. Although there is evidence of considerable volcanism during the Pleistocene, it is not obvious that such an event would have been sufficient in itself to trigger an ice age. Additionally, geophysicists Maurice Ewing and W. Donn have stressed the importance of greater-than-normal snowfall in building both Alpine and continental glaciers, and emphasized this relationship to positions and temperatures of oceanic currents. Earth's oceanic cold-water currents—notably the Labrador and Chilean—chill neighboring high-elevation lands, creating the potential for limited glacier formation. In conjunction with these effects, the cogent observations of Serbian astronomer-mathematician Milutin Milankovitch (1879–1958) may be especially relevant to Pleistocene continental glaciation. He proposed that three characteristics of Earth movement produced astronomical cycles and perturbations that best explain repeated alternations of glacial and interglacial periods during the Pleistocene:

1. The Earth's orbit is elliptical, causing our distance from the sun (and the energy we receive) to vary by approximately 2% over the course of a year.
2. The tilt of the Earth's axis varies from 21–24° over a 40,000-year period; currently the tilt is 23 ½°.
3. Another superimposed "wobble" of the Earth's axis occurs over a 26,000-year cycle; this effect is termed *precession*.

This trio of variations act together to modify the amount of heat received from our sun. Some glaciologists and climatologists predict worldwide temperature excursions from normal averages every 20,000 and every 100,000 years. When the short-term and long-term cycles coincide, climate change would be most severe. Most likely, several causes operating in tandem have created past frigid periods.

Continental Ice—Destructor

In its advancing mode a continental glacier acts primarily as a destructor. Initially, mountain (Alpine) and piedmont glaciers accumulate in elevated semi-circular depressions (cirques) and elongate basins. As these expand and spill over, the enlarged ice lobes become rivers of ice, collecting loose rock and soil from valley walls and floors. This abrasive mix, like sandpaper, abrades and sculpts the existing landscape. Eventually the isolated ice masses coalesce, creating a continuous ice blanket. With lowering temperatures and increasing accumulation of snowfall, a thick, creeping, conveyer-beltlike ice mass continues to assimilate rock and soil ingredients. These grinding tools pluck away at loose rock while polishing and grooving immovable bedrock. Underlying residual soils—the end products of the decomposition of exposed bedrock—are stripped away and incorporated within the engulfing ice or "bulldozed" in front of it. In Quebec and the Adirondack Mountains, spotty, thin, *transported* soils and extensive exposures of bedrock attest to the incorporation of pre-glacial residual soils and deposition elsewhere.

Now that eastern New York was completely ice-laden, the Hudson Valley was plugged and the Proto-Hudson River was either nonexistent or hidden under the ice. Beneath the ice were the beveled-off highlands (the Catskill Mountains and Adirondacks), and deepened and widened lowlands (the Taconic Hills)—a flattened and smoothed topography compared to the pre-glacial scene.

The tremendous weight of the continental ice exerted an overburdening stress on the buried topography. This caused the rock crust to sag—more so at the point of ice origin (northern Quebec), and less at the southern terminus of the ice (at the present-day location of Long Island). In nearby unglaciated regions south of the ice terminus, compensatory (isostatic) uplifts occurred. Connally and Sirkin (1986) have shown that readjusting present shorelines of glacial Lake Albany (discussed later) to water-level horizontality reveals that crustal rebound has been about 20 feet in southern Dutchess County near Beacon and about 80 feet in southern Columbia County near Livingston, a distance range of 41 miles. Can we predict the amount of crustal rebound in Quebec? We may safely assume that the ice blanket was thickest at its source, and therefore should exert the greatest stress on the rock crust. The Canadian Shield consists of very compact, intensely metamorphosed rocks, however, whereas those of the Hudson Valley are less compact sedimentary rocks; the latter should buckle more than the former. Taking these factors into consideration, crustal rebound in northern Quebec may be conservatively estimated at approximately 250 feet. Quebec's topography was probably largely below sea level during glacial maximum.

What prevented the Pleistocene ice from advancing further south? Some recent research suggests an answer. The study of ice cores obtained in Antarctica and Greenland reveals bubbles of trapped ancient air. In 25,000-year-old ice, these bubbles contain levels of carbon dioxide that are greater than current levels. Where did this additional carbon dioxide originate, and did this increase in carbon dioxide initiate the waning of Pleistocene ice?

A clue to this mystery may be the release of methane *clathrates* in continental shelf sediments. *Clathrates* are compounds formed by the inclusion of molecules of one kind in spaces within the crystal lattice of another. Once liberated into the atmosphere, methane oxidizes into carbon dioxide and water. Carbon dioxide is a "greenhouse gas" that absorbs heat radiation escaping from the Earth's surface; increases in atmospheric carbon dioxide could raise global temperatures, resulting in increasing glacial melting.

Glacial melting also results when the glacier advances into warmer latitudes. When the leading edge of the glacier experiences substantial melting, there are two possible results—the ice front may become *stationary*, or it may *retreat*. As ice flows slowly forward, it accumulates a cargo of variously sized particles—from house-sized down to clay. Even during cold climate cycles, the glacier may extend far enough south so that warmer temperatures will melt the ice front at the same rate as it pushes forward; the front will then be stationary. As this occurs, the material the ice carries is deposited at the front of the glacier, something known as a glacial *moraine*. Long Island is a composite of two such deposits. The Ronkonkoma Moraine comprises the southern "fluke" of the island; a later-forming stationary ice front produced the Harbor Hill Moraine comprising the northern "fluke."

Continental Ice—Constructor

When melting of the ice front exceeds the rate of forward movement, the glacier is considered to be "retreating." In this mode a continental glacier acts primarily as a constructor. As the climate warmed, blocks of ice calving at the ice front drifted into ponding meltwater. Most of these transient lakes have dried up or left smaller relicts. Notable ones that have endured are Lake Champlain and the approximately thirteen Finger Lakes of central New York; these elongated, U-shaped lake basins were formerly occupied by narrower V-shaped river valleys and their tributaries. The five Great Lakes are also glacial in origin, and in the past drained eastward through the Mohawk and Hudson Valleys.

In eastern New York the largest extinct lake, termed *Lake Albany* by geologists, flooded the Hudson Valley (Fig. 115). With a shape similar to present Lake Champlain, but much more extensive, it stretched from the Lake George area southward to southern Dutchess County, and extended from the base of the Helderberg Escarpment on the west to the vicinity of Chatham on the east. Saratoga, Ballston, Round, Nassau, and Kinderhook lakes are vestiges of Lake Albany. This ancient waterbody and its subsidiaries existed for about 5,000 years, based on radiocarbon dating of plant remains. Fossils, however, are rare in these lake deposits; frigid water, the short life span of the lakes, abnormal turbidity, and other factors made the environment hostile to their formation.

How rapidly did the ice front melt back to the north? Using radiocarbon dates of pollen in the earliest post-glacial sediments, Connally and Sirkin (1986) demonstrated that the continental ice (at its maximum in the Long Island area around 21,750 ybp [years before present]) receded from the Shenandoah Moraine (a glacial deposit located east of Beacon in southern Dutchess County) from 17,950 ybp to the Poughkeepsie Moraine at 17,210 ybp. The ice continued to retreat to the Red Hook Moraine (in northern Dutchess County) by 16,700 ybp. Thus, 34 miles of recession took place over approximately 1,250 years—about 50 yards (half a football field) a year. By 13,800 ybp, the Albany area was empty of continental ice.

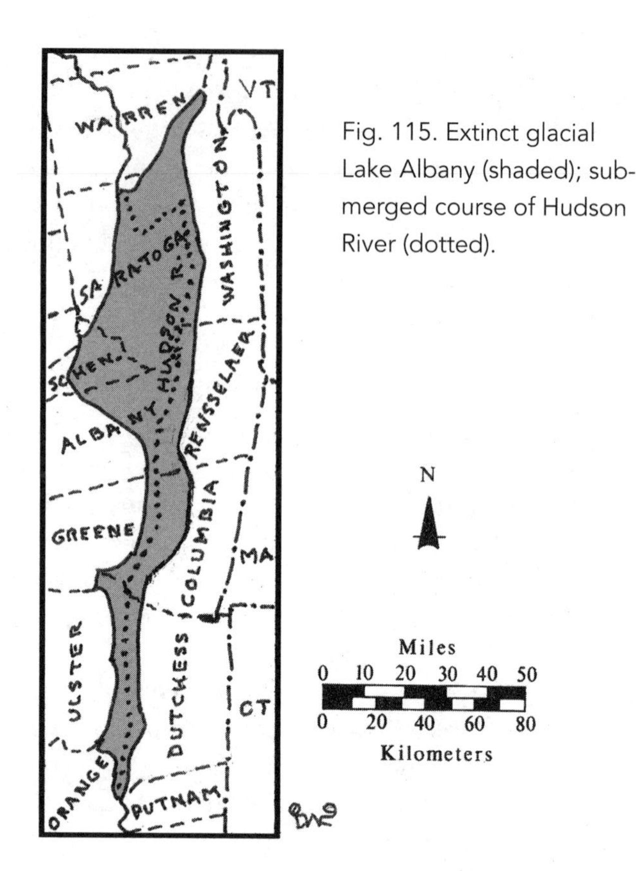

Fig. 115. Extinct glacial Lake Albany (shaded); submerged course of Hudson River (dotted).

Eventually, "spring" returned to New York State. Climatic warming brought about a gradual but intermittent northward retreat of the Pleistocene ice blanket. The removal of ice from the redesigned landscape allowed lichens, mosses, grasses, and subsequently, shrubs and conifer trees (cedars, firs, pines, and spruces) to recarpet the land relatively quickly. Because food was readily available, plant-eating animals could repopulate and be sustained on the renovated surface. The forerunners were grazers such as the barren-ground caribou (Fig. 116), wooly mammoths (Fig. 117), musk oxen, and their predators—the dire wolves (Fig. 118). The wooly rhinoceros was an inhabitant of Eurasia. Prolonged warming encouraged the encroachment of flowering and berry plants, along with hardwood trees (elms, maples, and oaks). Entrenchment of these diverse plants ultimately spawned flowering meadows and forests. These supermarket habitats made it possible for the existence of diverse animals with discrete dietary habits. Among them were the browsers (bark-eaters, berry-eaters, and branch-eaters) such as the woodland caribou, moose-elk (*Cervalces*) (Fig. 119),

Fig. 116. Barren Ground Caribou (*Rangifer arcticus*), with human silhouette for scale. Painting by Robin Rothman. Printed with permission of the New York State Museum, Albany, NY.

Fig. 117. Wooly Mammoth (*Mammuthus primigenius*) with human silhouette for scale. Painting by Robin Rothman. Printed with permission of the New York State Museum, Albany, NY.

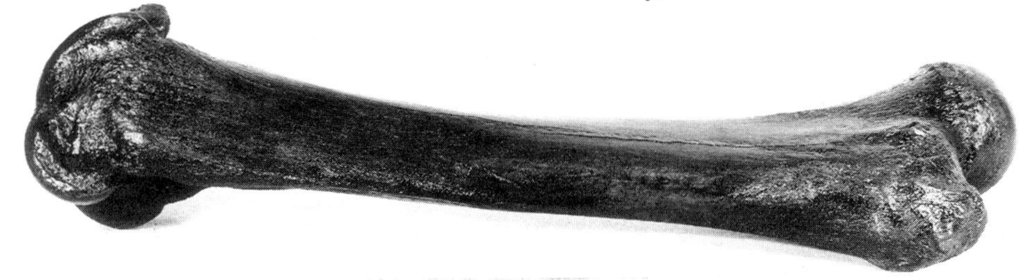

Fig. 122. Mammoth femur (thigh bone), approximately 4 feet long. Kitchawan, Westchester County, .3 miles west of Mt. Kisco. Collected by the author with Edgar M. Reilly, Jr. Printed with permission of the New York State Museum, Albany, NY.

giant bears, mastodons (Fig. 31, 32, 33), and 6-foot-long beavers. New grazer immigrants included giant longhorn bison (Fig. 120), peccaries (Fig.121), and the Jefferson (Columbian) Mammoth (Fig. 122)—a warm-weather cousin of the wooly mammoth. The bizarre 10-foot-tall ground sloth was an ungainly visitor. Eventually some of these residents migrated northward or southward. Others evolved to become smaller, and some became extinct.

Accompanying this zoo of vegetarians were their predators—including humans. Human occupation in the Hudson Valley dates from about 12,650 years ago. Evidence for this comes in the form of a Clovis fluted javelin or spear point (Funk, et al., 1970) from a peat bog in Orange County. This artifact was discovered amidst caribou bones that were radiocarbon-dated as 10,580 B.C. ± 370 years old. Mastodon and moose-elk bones from the same area (but not the same site) were radiocarbon-dated at 8,050 B.C. ± 160 years old, and 9,000 B.C. ± 150 years old, respectively.

Tributary streams deposited gravels, sands, silts, and clays (Fig. 123) into differing sections of Lake Albany. In Columbia County, for example, clays (used in brick making) were deposited in the Stuyvesant–Newton Hook–Hudson region, but gravels and sands (which provide excellent aquifers) provide a public water supply at Kinderhook Village and Schenectady. Where non-porous clays abound, water scarcity problems exist. In Albany, over 175 vertical feet of alternating bands of light and dark clays (termed *varves*) were encountered while pile driving to bedrock during construction of the Empire State Plaza. Each pair of bands documents one year of deposition. By counting these bands it is possible to calculate the number of years during which the clay sequence accumulated.

Clay deposits are unstable and produce hillside slumping—a problem for engineers designing roads, foundations, and buildings. As an example, several years ago US 9 south of Stockport Creek was relocated because of repeated slumping along the old highway;

Fig. 118. Ring of musk oxen (*Ovibos moschatus*) being attacked by Dire Wolves (*Canis dirus*), with human silhouette for scale. Painting by Robin Rothman. Printed with permission of the New York State Museum, Albany, NY.

Fig. 119. Moose-Elk (*Cervalces scotti*), sometimes referred to as Stag Moose. Painting by Wayne Trimm. Printed with permission of the New York State Museum, Albany, NY.

the weight of additional blacktop to maintain pavement level only aggravated the situation.

With the continued lowering and eventual disappearance of Lake Albany, fast-flowing meltwater from lingering upland glaciers cascaded off the partly covered Taconic, Catskill, and Adirondack mountains into the reborn Hudson River. This artery to the Atlantic was laden with detritus ranging in size from clay to boulders.

New York State presents a "dreamscape" for those scientists who specialize in glacial geology. In addition to the previously described *moraines* and bedrock *striations*, other glacial relicts found today include **potholes** (Fig. 124), *drumlins*, **erratics, eskers** (Fig. 125), **kames, kettle hole lakes**, and *terraces*—new landforms for this area. While a complete description of these features is beyond the scope of this book, drumlins, erratics, and eskers deserve special mention.

Drumlins are elliptical-shaped hills oriented approximately north to south. From the air they resemble the back of a whale breaching the surface of the ocean. It is thought that drumlins were created by the re-advance of glacial ice over an earlier-formed moraine. Many show a steeper north slope, indicating that the flow of overriding ice was from the north. A

Fig. 120. Long Horned giant Bison (*Bison latifrons*). Painting by Robin Rothman. Printed with permission of the New York State Museum, Albany, NY.

Fig. 121. Group of flat-headed pig-like peccaries (*Platygonus compressus*); height at shoulders was 28 inches. Painting by Robin Rothman. Printed with permission of the New York State Museum, Albany, NY.

Fig. 123. Stratified sand, silt, and pebbly gravel with later Pleistocene channel fill, Dutchess County. Professor Scott Warthin of Vassar College for scale.

few have bedrock cores. The largest drumlin field in the world lies between Syracuse and Rochester, north of the major Finger Lakes. It is exquisitely displayed along the New York State Thruway (I-90).

Erratics are boulders that have been transported by ice movement to new locations; typically they don't match the bedrock beneath them. Along the north side of Columbia County 24, at the hamlet of Red Rock, there is a splendid example of such an erratic (Fig. 126). This 25-foot-diameter, reddish-white quartzite is an identical match for the Monkton Quartzite—originating at least 50 miles to the north-northeast! Along with other such mega-boulder erratics in eastern New York, other transported boulders, cobbles, and pebbles of "stone foreigners" represent samples conveyed from Quebec and northern New York.

Locally, glacial debris may completely plug preglacial watercourses. While drilling for foundation preparation for the Colonie Center Mall (between Schenectady and Albany), a filled channel of the Proto-Hudson River was uncovered. The blockage of the "Colonie Channel" forced the Hudson River to shift to its present course further to the east. The stream and river drainage network that we see today in New York State was imprinted upon sediment-filled depressions and uncovered bedrock of the glacially rounded and lowered hills of the southern Adirondack, Catskill, and Taconic mountains.

Since the glacial meltdown, the Earth has experienced a general trend of global warming, perhaps accelerated by modern human activities coupled with apathy toward protection of our environments. What

Fig. 124. Large glacial pothole in Lower Ordovician Chuctanunda Creek Dolostone, with Middle Ordovician Trenton Limestone unconformably atop in the background, along Canajoharie Creek at southern end of the village. Betty Fisher for scale.

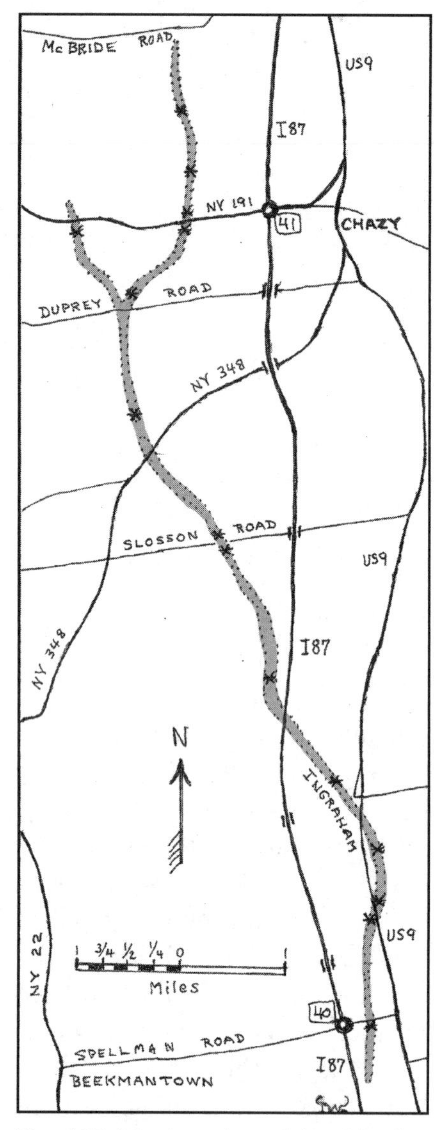

Fig. 125. The Ingraham Esker (shaded)—the best example of this type of recessional glacial deposit in New York State. Northeastern Clinton County, between Chazy on the north and Plattsburgh on the south. Observed clam and snail sites are indicated by asterisks.

would happen if all 25 million km^3 of ice presently existing on Earth suddenly melted? Ocean levels would rise about 200–250 feet, effectively flooding most coastal cities of the world. The Great Lakes would be enlarged to the extent of appearing as a single inland sea. Niagara Falls would be submerged. Continents would be noticeably smaller, and Long Island and Florida would disappear beneath the surface of the ocean. The Manhattan skyline would become a series of isolated, water-locked pillars. The Hudson River would assume Amazonian proportions, and all riverfront cities and villages would be submerged. Portions of the Mississippi Valley might broaden to a seaway connecting the Gulf of Mexico to the Arctic Ocean—a situation that existed during the Late Cretaceous Period.

Conversely, if renewed ice sheets overspread more than half of the continents, sea level might drop 400–500 feet, extending the shorelines far out upon the continental shelves. Land areas would undergo significant enlargement. An unglaciated Bering bridge would once again connect Eurasia with North America, allowing renewed interchange of animal life. An Indonesian land bridge would surface, permitting migration between Southeast Asia and Australia. This is not beyond probability, because we know that Pleistocene shorelines reached far out onto the continental shelves. For example, the Hudson Bay, Baltic, and North Sea regions hosted treeless tundra environments and featured ground-loving plants. Along the eastern North American coast, human artifacts and other terrestrial animals and plants have been recovered in up to 250 feet of the Atlantic Ocean, attesting to previously broader coastal plains. Because elephants are unsuited for oceanic swimming, the presence of mammoth and mastodon bones at these depths provides proof that ocean levels during the Pleistocene were considerably lower than they are today.

Neither extreme condition ("meltdown" or "freeze-up") will happen in our lifetime. Many glacial geologists predict that global cooling might commence at approximately 2300 A.D. and reach maximum ice coverage between 3000–3500 A.D.—probably persisting to 5000 A.D. There will be a profound plant elimination, coupled with mass animal mortality and/or migration southward. Surface configurations will change dramatically as the entire cycle of glacial destruction and construction is repeated. Once again, the Earth will be in the grip of another "long winter."

Fig. 126. Glacial erratic of Lower Cambrian Monkton Quartzite between Indian Creek and north side of Columbia County 24, Red Rock, Columbia County. Monkton Quartzite is found in place near Burlington, Vermont.

part three

Economic Resources from the Ground in Columbia County

Chapter 30

Water, Water, Everywhere

Water is our most vital mineral resource. Without it, plants and animals could not exist. Hydrology is that branch of geology that deals with global water (liquid and solid) and its properties, circulation, and distribution on the Earth's surface from the moment of precipitation until it is returned to the atmosphere by evaporation, is discharged into bodies of water, or is percolated into the soil and rocks. For the past few decades hydrology's scope has been expanded to include environmental and economic aspects. Hydrologists (or geohydrologists) and hydraulic civil engineers are the scientists who specialize in these topics.

Surface Water

Ninety percent of Columbia County lies within the Hudson River drainage basin. Tides up to three vertical feet create tidal flats of muckland along the Hudson River, which is navigable to oceangoing vessels as far north as Albany. Most of Columbia County is drained by the primary tributary streams to the Hudson: Kinderhook and Claverack creeks and their confluence distributary Stockport Creek; and Roeliff Jansen Kill and its respective secondary tributaries. Queechy Lake Brook, with its headwaters in Queechy Lake, drains into the Stony Kill and thence into Kinderhook Creek. In the Canaan, State Line, and Egremont Quadrangles, the Green River (with headwaters in No Bottom Pond) and Flat Brook (with headwaters in Beebe Pond) empty into the Housatonic drainage system in western Massachusetts and Connecticut. Among the natural lakes and ponds within the county, the following are the largest and available for boating, fishing, and swimming: Kinderhook Lake on the Kinderhook Quadrangle; Queechy Lake on the Canaan Quadrangle; Lake Taghkanic (formerly Lake Charlotte) on the Ancram Quadrangle; Copake Lake on the Hillsdale Quadrangle; and Robinson Pond on the Copake Quadrangle. Artificial water bodies (reservoirs) for public water storage exist at Philmont and at Becraft Mountain, where an abandoned limestone quarry holds additional water for the city of Hudson.

A spring is a place where groundwater flows naturally from rock or soil onto the land or into a body of water. Its location depends on the nature and relationship of the rocks or sediments, their permeability, the position of the water table, and the character of the topography. Artesian water is groundwater under hydrostatic pressure, expressed in pounds/square inch at the land surface, or height in feet above the land surface. Springs dot the landscape throughout the county. Should they show a linear pattern, a concealed fault may be suspected because the underground water may be rising along a major fracture. Most of the county's springs yield 10–100 gallons/minute.

A most remarkable warm spring occurs near the east side of NY 22 in northern Lebanon Springs. Here the water is uniformly 73° F, about 20° warmer than the normal temperature of the groundwater in this area. This may reflect a deep-seated source of heat below, or surface water that has penetrated to considerable depths and returned to the surface along fractures. It is noteworthy that the spring is astride a major fault, with the late Proterozoic Everett Phyllite-Schist overriding the Cambrian-Ordovician Wappinger Group of dolostones and limestones in the valley. The Lebanon Spring yields 75 gallons/minute with a hardness of 50–150 ppm (parts per million of mineral matter). "Soft" water has a hardness of less than 50 ppm, whereas "hard" water has a hardness of 200–300 ppm and requires water softeners for efficient household use. Some springs in the County have a high hydrogen sulfide (H_2S) content, imparting a "rotten egg" odor to the water; this usually reflects pyrite-bearing shale beneath the soil cover. Other springs may be heavily charged with carbonic acid (H_2CO_3), emitting carbon dioxide gas; this usually denotes carbonate strata beneath the soil.

Underground Water

Groundwater exists sporadically in the zone of aeration above the water table, and universally in the zone of saturation below the water table. The water occurs in pores in sediments and rocks and in cracks in the latter. Above the water table, water is held by molecular attraction and is available for absorption by root systems of plants, but is in too short supply for human consumption. One must look to the sediments and rocks below the water table for satisfactory amounts of water for human use. Of course, neighboring surface water bodies are preferable, as the volume is virtually unlimited and initially less costly to acquire. Whether surface or subsurface water is intended for human use, one should be wary of possible corporate waste pollution or pesticide infiltration from neighboring farms.

The most productive aquifers are gravels and sands, owing to their greater porosity and increased permeability (capacity for liquid flow via interlocking pore and fracture connection). Almost equally desirable are loosely cemented sandstones and vuggy (containing hollows or cavities) or cavernous limestones and dolostones. Shales, tightly cemented sandstones, graywackes, and plutonic and metamorphic rocks customarily yield very little water, as they are relatively nonporous and usually impermeable. Should, however, these compact rocks be severely fractured by crosscutting joints or numerous faults, they may furnish adequate supplies for limited residential use.

As with surface springs, subsurface water may be tainted with bad-tasting, unhealthy, or even cancer-forming components. Some of these undesirables are hydrogen sulfide gas, other sulfur compounds, iron compounds, lead, methane, carbon dioxide, abnormally high concentrations of calcium, magnesium, phosphorus, radioactive elements, discharged organic pollutants from corporations, and pesticide contaminants from nearby farms. These may not only stain bathroom appliances and instill a bad taste to drinking water, but also could create serious digestive, laxative, or skin rash problems. In order to avoid such disorders, household water should be periodically tested for potentially harmful ingredients.

Dowsing

"Science is the great antidote to the poison of enthusiasm and superstition." Adam Smith, 1776

Dowsing is the practice of locating underground water, buried mineral deposits, or a variety of objects using a divining rod or forked wood (rhabdomancy), or a pendulum (pallomancy). Some dowsers may also claim to be able to diagnose diseases, determine the sex of unborn babies, or locate missing persons or dead bodies.

Dowsing can trace its ancestry back to the Bible (Numbers 20: 7-11), where Moses, using his brother Aaron's rod, "struck the rock twice, and water gushed out." In northwestern Africa, 6,000–8,000 year-old cave paintings depict a dowser at work. Ancient Persians and Medes used divining rods, but the Greek and Roman naturalists did not mention them—although they gave advice on where to find water using characteristics of the terrane. During the Middle Ages the dowsing concepts became contagious. An uneducated populace seeking relief from disease, poverty, and starvation appealed for solace from mystics who promised much but furnished scant comfort from the miseries of the "Dark Ages."

Until the past few decades, traditional dowser's tools were forked twigs of witch hazel, willow, or a fruitwood—especially fruits with pits. But today most dowsing instruments are made of synthetics such as metal coat hangers, metal rods, metal pendulums, and various plastics. Some dowsers never go out into the field; dangling pendulums over maps, they "locate" selected sites of water and other objects thousands of miles away!

Whenever dowsers are confronted with scientifically rigorous tests of their self-claimed abilities, they score no better than the average person. In reality, it's not hard to find water under the Earth's surface—provided you dig deep enough! Contrary to dowser's beliefs, groundwater is not restricted to veins, domes, or underground rivers or lakes. Instead, water is sporadic in the zone of aeration above the water table, but it is everywhere in the zone of saturation below the water table where it occupies pores and fractures in the rocks and the interstices in the sediments. Accordingly, water dowsers can never be wrong. Why, then, would anyone consult dowsers (or, for that matter, geologists) since digging anywhere will yield water eventually?

To people unschooled in the nature of water, dowsers tend to be more colorful and charismatic than staid geologists and hydraulic engineers. Geologists and engineers are scientists who uphold and use the laws of chemistry, geology, and physics in order to locate and evaluate underground water. Not only can these hydrologists locate water, but they can also ascertain its quality, approximate the quantity to be recovered, determine the safe removal rate and the resultant impact of that withdrawal on adjacent wells and streams. Additional hydrological data will include the chemical composition of the water, a geological map, and geological cross sections of the rocks and sediments. This is the scientific approach used by consulting civil engineers, hydrologic geologists, state geological and water resource agencies, and the U. S. Geological Survey. To date, no scientifically accepted studies sustain the validity of dowsing.

Chapter 31

Metal Mining

Iron Mining

The earliest iron manufacturing in New York State took place at Ancram, in Columbia County, on the Roeliff Jansen Kill (formerly Ancram Creek). Here, Robert R. Livingston founded and operated a smelter in 1748; it was demolished in 1854. From 1770–1790 he also operated the Maryburgh Foxes smelter at the New Forge on Taghkanic Creek; this was abandoned in 1796. During the late eighteenth and throughout the nineteenth centuries, iron mines in Columbia, Dutchess, Putnam, Westchester, and Orange counties supplied ores to smelters, which in turn furnished iron to foundries for steel manufacture. Columbia County's iron mining industry was centered in two geologically different regions: the Church's Hill (formerly Plass Hill)–Cedar Hill–Mt. Tom "highland" belt; and the Hillsdale–Copake–Ancram "lowland" belt. Noteworthy is the fact that the largest open-pit iron mine in the world during the 1880s was the Tilly Foster Mine, west-northwest of Brewster, Putnam County. Minerals from this classic site are highly prized by collectors. In 1888, New York State ranked third, behind Pennsylvania and Michigan, in iron ore production in the United States. Today, there are no active iron mines in the Empire State.

The western or "highland" belt extended from Church's Hill on the north to Mt. Tom (formerly called Mt. Thomas) on the south and lies entirely within the Hudson South topographic map at elevations from 350 to 518 feet. The ore-bearing unit (Germantown Formation) dips from 25–45° ENE. **Siderite, limonite**, and some **hematite** occurred in calcitic and dolomitic conglomerates, quartzitic sandstones, and (rarely) in silty shales. Iron-rich zones varied from a few feet to a maximum of 30 feet on Cedar Hill. All were open pits except at Mt. Tom, where workings went underground. Excavations at Church's Hill reached 50 feet deep; at Cedar Hill none exceeded 25 feet; at Livingston's "mine" east of the main Cedar Hill excavations, a vertical shaft bottomed at 40 feet.

The main producer, however, was the Burden Mine (named for Howard H. Burden, company president) on County Route 10 at Burden (Hudson South quadrangle). The principal owner was the Hudson River Ore and Iron Company, which assumed control in 1882. A 3.5-mile narrow-gauge railroad led to the Hudson River, where ten roasting kilns removed sulfur and carbon dioxide so as to reduce the iron concentrate preparatory to smelting. Then the roasted ore was shipped either to smelters in Troy, New York, Scranton, Pennsylvania, or Franklin Furnace, New Jersey, for further enrichment. Burden ore ranged from phosphorus-rich (for pig iron) to Bessemer-quality (for steel). Quartz and pyrite were common mineral associates. The mine operated from 1873–1901 and yielded slightly over one million tons of ore. Today, the underground Burden site is used by the Iron Mountain Security Corporation for the rented storage of records.

The eastern or "lowland" belt was in the Hillsdale–Copake–Ancram valley (Hillsdale, Egremont, Copake topographic maps). This iron-suite yielded ores that were primarily iron hydroxides (**goethite**, limonite), popularly referred to as bog iron ore or brown ochre. Occasionally, iron oxide (hematite) and iron carbonate (siderite), both brown, were found deeper in the diggings. This relationship implies that iron ores above the water table were transformed hydroxyl iron oxides or oxidized iron carbonates. All of the iron deposits occurred sporadically along an ancient erosional surface near and at the summit of whichever formational unit capped the Wappinger Group of limestones, dolostones, and admixtures of these.

At least seven former "lowland" belt iron ore workings are known:

Copake Mine—(in Wappinger Group, Briarcliff Formation)—Copake Quadrangle, northeast of the abandoned depot on the former Harlem Division of the New York Central Railroad, an open pit (now the adult pool in Taconic State Park) 500′ in a north-south direction and 200′ in a west-east direction, rock dips 35–40° ESE, last worked in August 1888.

Haight or Foster Mine—(in Wappinger Group, Rochdale Formation)—Hillsdale Quadrangle, one-quarter mile east of North Hillsdale, in valley near Walloomsac black phyllite on the west, opened in 1862, idle since 1885.

Hillsdale or Smith Mine—(in Wappinger Group, Briarcliff Formation)—Egremont Quadrangle, 3 miles east of Hillsdale near the state line, ore was carried by wagon to the former railroad depot at Hillsdale, mine opened in 1834, abandoned in 1887.

Mitchell Mine—(in Wappinger group, Briarcliff Formation)—Egremont Quadrangle, 3 miles NE of Hillsdale, opened in 1800, idle in 1877.

Morgan or MacArthur Place Mine—(Wappinger Group, Copake Formation)—Copake Quadrangle, 1.9 miles SE of Ancramdale, opened in 1776 by the Livingstons (one of the oldest mines in Columbia County), reported to yield 48% iron, abandoned in early 1890s.

Reynolds Mine—(in Wappinger Group, Rochdale Formation)—Copake Quadrangle, 2.2 miles northeast of Ancramdale and 0.4 miles east of East Ancram Road, opened in 1857, idle since 1887

Weed Mine—(in Wappinger Group, Pine Plains Formation)—Copake Quadrangle, 2 miles south-southeast of Copake and east of NY 22, narrow-gauge railroad carried ore to Harlem Division of former New York Central Railroad, ore body 1,600′ long with average width 70′ and depth 120′, yielded 43% iron, 100,000 tons removed by 1889, opened in 1840, abandoned in 1901.

Differing explanations have been offered for the age and origin of the Columbia County iron ores. Dana (1884) was the first to specifically address the issue by attributing the accumulations to initial deposits in sedimentary basins. Eckel (1905) disagreed, claiming they were replacements of limestone (calcium carbonate) by iron carbonate. Smock (1889) provided the most comprehensive description of the Taconic iron ores, but prudently sidestepped questions of origin and age. Ruedemann (1931) and Newland (1936) both subscribed to a contemporaneous origin with the enclosing rock. The former assigned a Normanskill (Early Medial Ordovician) age, whereas the latter assigned an Early Cambrian age. Contrary to Ruedemann's belief that the western "highland" belt of iron ores lay in the Normanskill Formation, they actually existed within the limestones, conglomerates, quartzitic sandstones, and silty shales of the older Germantown Formation (Early Cambrian through Early Ordovician).

At present, the derivation of the Taconic iron ores remains a baffling and equivocal geological problem. An alternative variation of previous interpretations is that the iron-bearing units originated as **magnetite** sands on ancient beaches and along fringing reefs and became concentrated by tidal and/or current action. The magnetite (ferric oxide, Fe_3O_4) was then converted, underwater, to siderite (ferrous carbonate, $FeCO_3$). Subsequent uplift, coupled with subaerial erosion, altered the siderite to hematite (ferrous oxide, Fe_2O_3) and goethite-limonite (hydrated iron oxides, FeOOH). This resulted in iron-rich residual soils spottily distributed on the Early Ordovician ancient landscape (unconformity). Because of differential erosion, a patchy distribution of iron deposits accumulated in erosional troughs, joint fractures, bedding planes, and solution cavities on a **karst** surface in a limestone-dolostone terrane. In the Mohawk Valley I have observed identical iron hydroxide fillings atop the Early Ordovician Tribes Hill Formation and Amsterdam Limestone—in the same Pre-Trenton erosional surface (unconformity) that caps the Wappinger Group.

When, in the late nineteenth century, abundant and richer magnetite lodes were discovered in the Adirondack Mountains (specifically in the

Port Henry–Witherbee area of Essex County and the Star Lake region of St. Lawrence County), the leaner siderite-limonite ores of the Hudson Valley were doomed. The heyday of Adirondack iron mining was also temporary, however. Destined for abandonment owing to their dangerous and expensive underground operations, they were supplanted in the 1950s by the open-pit, cheaper-to-remove, hematite ores of northern Minnesota. Today, it is difficult to locate relicts of a once-thriving iron mining industry in Columbia County. Wholly or partly filled-in and plant-overgrown excavations—and abundant quantities of conchoidal-fracturing, often lustrous blue-grey-black slag (the residue from roasting ovens or smelters)—are about all that survive of an industry that "rusted away."

Lead Mining

Columbia County's sole operational lead mine was the Ancram (or Livingston) Mine (Fig. 127). It was first reported in a scientific journal by Charles A. Lee (1824), who stated that it operated for four to five years and was abandoned in 1823 owing to scarcity of ore. In addition to lead "sulphuret" (galena), Lee mentioned zinc "sulphuret" (sphalerite), molybdate of lead (wulfenite), pyritous copper (chalcopyrite), green carbonate of copper (malachite), sulphate of Barytes (barite), "sulphuret" of iron (pyrite), quartz, milky quartz, radiated quartz (smoky quartz), puddingstone (conglomerate), and clay. He further noted that the cubic galena was "argentiferous" (silver-bearing).

Lewis C. Beck, mineralogist of the first New York State Geological Survey, reported (in *Mineralogy of New York*, 1842) two or three veins of galena (lead sulphide, PbS) observable at the surface with crystalline and granular galena disseminated in calcite and quartz. Beck described the mine to be a "vertical shaft 70 feet deep and from this extended levels which communicated with another shaft, now used as a ventilator." He goes on to state, "The mine seems at present to be judiciously worked, although I have reason to believe that it has not answered the expectations of the proprietors." This comment indicates that the mine was reopened after Lee's visit.

The 10-foot-diameter vertical shaft is in the Pine Plains Dolostone, with a major reverse fault surfacing about 80 feet to the northwest; the Walloomsac black phyllite crops out another 25 feet beyond. At the time of this author's visit, 30 feet of rock wall was visible above the water-filled shaft. Mather analyzed the sphalerite (zinc sulphide, ZnS) from the mine and found that although the ore yielded 118 ounces of silver per ton of rock, "The difficulty of separating the quartz from the galena renders the processes necessary for its reduction comparatively tedious and expensive."

The site in question is on a gentle southeast slope 0.15 mile southeast of Ancramdale (formerly named Ancram Lead Mines on old maps) on the northwest side of NY 82 on the Copake 7 ½-minute topographic map. The fenced-in adit of the vertical shaft is in a wooded area. Vegetation-covered dump rocks show that the galena occupies micro-fractures in the dolostone. On the hanging wall side of the major reverse fault, the Pine Plains Dolostone dips 55–65°E. Elsewhere in the county, small flecks of galena and sphalerite sporadically occur within the Wappinger Group carbonates in the Canaan, State Line, Hillsdale, Egremont, Copake, and Ancram Quadrangles. No prospects that this author has seen would warrant future profitable development.

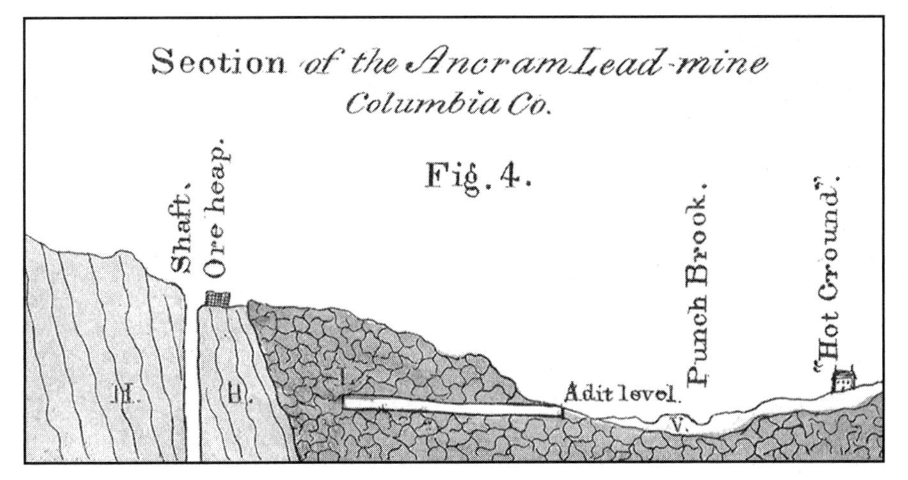

Fig. 127. A section of the Ancram Lead Mine, Columbia County. This hand-colored plate appeared in Mather's *Geology of New York, Part I, Geology of the First Geological District* (1843).

Chapter 32

Limestone

One of the most diversified geological economic resources is the calcium carbonate-rich sedimentary rock, limestone. It constitutes highway and driveway aggregate, furnishes fertilizer in the form of agricultural lime, has been used for building stone, large blocks make excellent riprap for holding unstable slopes, and fossiliferous types are educational conversation pieces for rock gardens. In the Hudson Valley, limestone has been, still is, and will continue to be the prime item for the manufacture of Portland cement (though annual production figures for counties are confidential).

Within Columbia County the purest limestone occurs at Becraft Mountain and at Mt. Ida (formerly termed Mt. Bob), the former in the northeastern Hudson South Quadrangle and the latter in the southwestern Stottville Quadrangle. At the northern end of Becraft Mountain, south of the Hudson Cedar Park Cemetery, a large, active limestone quarry (Colarusso and Sons) furnishes highway and driveway aggregate. As of this writing, the New Scotland Limestone is being quarried. Their inactive quarries east of Newman Road are in the Manlius, Coeymans, Kalkberg, and the New Scotland Limestones. Other inactive quarries were the source for cement rock for the defunct Lone Star (east side of Newman Road) and Universal-Atlas (west side of Newman Road) cement corporations. These companies blended all of the Helderberg Group Limestones. The Saint Lawrence Cement Company, currently non-operational in Columbia County, holds ownership to a former Universal-Atlas quarry just east of US 9.

Mt. Ida is a low hill southwest of the intersection of NY 66 and NY 9H and the site of the long-abandoned limestone quarry that is now the property of Colarusso and Sons. It yielded Manlius, Coeymans, and Kalkberg Limestones. Future quarrying there would not be cost-effective, as reserves are very small. Structurally, the site is of geological importance as it displays two episodes of folding separated by an angular unconformity—Early Devonian Manlius Limestone resting upon Early Ordovician Stuyvesant Falls shale.

Chapter 33

Cement and Concrete

Cement and concrete, among our most important building materials, are two words that are frequently misused. Cement is a manufactured grey powder. When mixed with water, cement creates a plastic mass that will harden or "set." Some sedimentary rocks (termed "waterlimes") are natural hydraulic cements because they possess the required ratio of carbonate rock (dolostone or limestone) and shale minerals that, when mixed with water, cause the sludge to harden. Notable New York natural cement rocks are the Late Silurian Rondout waterlime (Rosendale and Whiteport Dolostone Members) in the Kingston region of Ulster County and the Howe's Cave region of Schoharie County, and the Late Silurian Bertie waterlime (Scajaquada and Williamsville Dolostone Members) of central and western New York. The Bertie Formation was used in the construction of the Erie Canal. Natural cement sediment is marl from Onondaga County, which also was used in canal construction.

Virtually all cement in use today is Portland cement, invented and named by Joseph Aspdin, a British bricklayer, in 1824. The name was chosen for its resemblance to Portland Limestone, a building stone widely quarried on the south coast of Great Britain. Portland cement is an adhesive produced by fine-grinding a carefully proportioned mixture of limestone, shale, silica, and gypsum, heating the mixture to incipient fusion in a rotary kiln, and fine-grinding the resulting clinker to a powder. In the United States, the Portland cement industry dates from 1875.

Concrete, on the other hand, is a mixture of cement (Portland or, rarely, natural), water, and aggregate. The aggregate may be sand or gravel, crushed stone (whatever is readily available), or cinders. Many people erroneously refer to cement sidewalks, cement highways, or cement bridges. Actually, they are composed of concrete. Concrete is durable, fireproof, watertight, and easily shaped when initially mixed. It requires little maintenance, but is prone to deterioration in climates where salt is applied to sidewalks, highways, and bridge abutments. Recently it has been found that coating with certain plastics retards this disintegration. Because of its propitious qualities, concrete is extensively used in construction of foundations, buildings, skyscrapers, walls, dams, bridges, piers, ramps, roads, and sidewalks. Where additional strengthening is essential, the inclusion of steel bars, beams, panels, and meshwork allows reinforced, precast, or prestressed concrete and concrete blocks to be used.

The following is a generalized flow chart that diagrams the manufacture of Portland cement by the various cement corporations (with slight modifications) in the Hudson Valley, including the defunct Universal-Atlas and Lone Star corporations at Hudson.

Fig. 128. The "alphabet" of Portland cement manufacture.

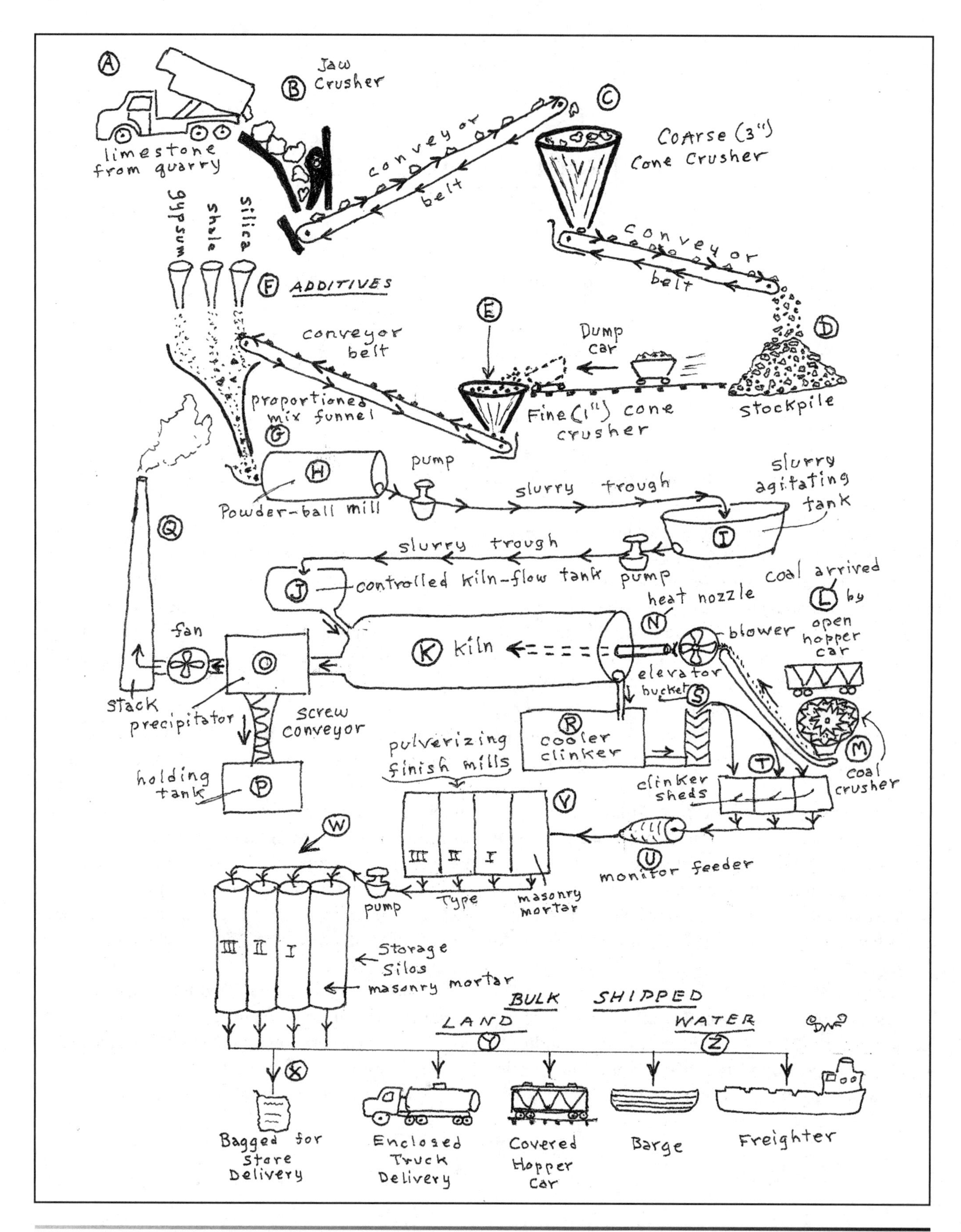

The "Alphabet" of Portland Cement Manufacture (Fig. 128)

A. Large blocks (up to two tons) of LIMESTONE are TRUCKED from nearby quarry to cement-making facility.

B. They are dumped into a huge pressure-driven JAW CRUSHER, which reduces the limestone blocks to no larger than football size.

C. Via conveyer belt, the crushed rock moves to a COARSE-CONE CRUSHER, which further reduces the limestone to no larger than three inches.

D. This is conveyed (sometimes over one mile) to STOCKPILES near the cement plant. The piles are categorized according to purity of limestone. Red shale, silica (sand), and gypsum are trucked in and stockpiled separately for late blending.

E. Via ore cars, trucks, or front-end loaders, stockpile material is separately run through a FINE-CONE CRUSHER and reduced to no larger than one-inch size.

F. SEPARATES are STORED in individual bins prior to blending.

G. They are blended in a PROPORTIONAL–MIX FUNNEL.

H. After this they pass into a POWDER–BALL MILL where water is added.

I. The resultant slurry is pumped via a slurry trough to a SLURRY AGITATING TANK(S).

J. It is then pumped to a CONTROLLED KILN–FLOW TANK.

K. The regulated slurry is then piped into the KILN, a 400′+ long cylinder (longer than a football field!) 17 ½′–20′ in diameter. Here, the slurry flows slowly through a section of steel chains that function as heat exchangers and heat-shield curtains. The exit end of the kiln is lined for about 100′ with firebrick, whereas the remainder of the kiln is lined with fire-retardant cement termed Gunite. This is fastened to the kiln interior with thousands of v-shaped stainless-steel prongs termed "hairpins." In modern cement plants, Gunite is sprayed on in layers and built up to about one foot thick. It is exceedingly hard and difficult to remove by jackhammer when it becomes necessary to repatch the interior.

L. In order to generate high temperatures within the kiln, BITUMINOUS COAL arrives by open rail hopper cars and is stored on a siding for quick availability.

M. When needed, the coal is fed to a COAL CRUSHER where it is pulverized.

N. It is then tunnel-blown through a HEAT NOZZLE into the kiln. Modern cement plants retain one or more backup fuels such as #6 heating oil or natural gas—or use these exclusively. The resultant flame shoots far into the kiln and, at temperatures of 2700–3000o F, effectively burns off all impurities, creating an extremely high-temperature viscous material, like molten lava. When operating at full capacity, 20–30 tons of coal per hour may be consumed.

O. The PRECIPITATION BUILDING, at the entrance end of the kiln, consists of a series of compartments with curtains of positively charged coil wires alternating with negatively charged metal panels, and placed five inches apart. The panels collect the dust originating from the kiln. At the bottom of the panels are shock rods that are hit by alternating "hammers" (wheels on a hinge), causing the dust to fall on hoppers below.

P. This is removed by screw conveyers to HOLDING TANK(S), later to be hauled away by truck.

Q. A tunnel-fan blows the residual steam from the precipitator and kiln to a STACK. If the precipitator is operated efficiently, only steam (not dust!) should exit from the stack. An 18–24" gap between the stack top and exiting steam denotes proper functioning of the precipitator.

R. The CLINKER–COOLER accepts the hot, molten residue from the kiln. Cool air is force-blown through small holes (¼") in moving stainless steel grates, which keep the now-formed clinker moving.

S. The clinker goes to a BUCKET ELEVATOR that reaches to the top of the building, allowing further air-cooling of the clinker.

T. Nevertheless, some still-hot clinker, now in balls with a maximum diameter of three inches, is dropped outside to CLINKER SHEDS where it is stored.

U. Stored clinker is then inserted into a MONITOR FEEDER, where clinker and gypsum are carefully proportioned.

V. They are then fed into FINISHING MILLS. Here, the mixture is milled to the type of cement desired. The higher the number, the more finely milled and the more stronger-setting the cement:

Masonry Mortar—the cheapest and least binding; for brick laying

Type I—general purpose: house foundations, sidewalks, septic tanks

Type II—highways, bridges, ship hulls, ramps, and piers

Type III—super structures; skyscrapers, earthquake-prone areas

W. The milled cement is then conveyed to STORAGE SILOS.

X. It is then BAGGED for retail store delivery.

Y. It is also BULK-SHIPPED BY LAND via enclosed truck or covered hopper railcar.

Z. And it may be BULK-SHIPPED BY WATER via barge or freighter (overseas).

Chapter 34

Bricks, Gravel & Sand, And Other Potential Economic Resources

Bricks—Clay Building Blocks

During the nineteenth cenury and the early half of the twentieth century, when bricks were the preferential facing for buildings, a robust industry thrived in the Hudson Valley. Many brickyards formerly operated along the east side of the river in Washington, Rensselaer, Columbia, Dutchess, Putnam, and Westchester counties, and along the west side of the river in Saratoga, Albany, Greene, Ulster, Orange, and Rockland counties. The low cost of transportation (Hudson River barges) and a bountiful market (New York City) were contributory factors toward a healthy economy. Because Hudson Valley clays held the proper proportion of clay minerals and the fluxing ingredients (iron, limestone, and dolostone), they were especially suited for the manufacture of bricks. These clays were deposited up to 250 feet above sea level in ancient glacial Lake Albany.

In Columbia County, once-profitable brickyards existed north of Hudson (Greenport Brick Corporation), about 1.5 miles northwest of Columbiaville (Empire Brick and Supply Company, Stockport Works), and in the neighborhood of Newton Hook (Carey Brick Company, Stuyvesant Works). All ceased operations prior to World War II. Mather reported (1843) that in 1837 Columbia County brickyards made 4,900,000 bricks out of the 130,860,000 made in the entire Hudson Valley, and that common bricks sold for $5.50/thousand and stock bricks sold for $20/thousand. The last surviving brickmaker in the Hudson Valley was Powell & Minnock Brickworks, Inc. near Coeymans in southeasternmost Albany County.

Gravel & Sand—Constructional Sediments

"The loose stones found on the sides of hills, and on the bottoms of valleys, when traced back to their original place, point out ... the great changes which have happened since the commencement of their journey." John Playfair, 1802

For economic purposes, gravel and sand are treated together because they are genetically related sediments in eastern New York. Gravel is an unconsolidated natural accumulation of rounded rock fragments resulting from stream, glacial, or wind erosion, and illustrating extensive transportation. The constituents are larger than sand (greater than 1⁄12 inch) and size-sorted as follows:

granules	1⁄12–1⁄6 of an inch
pebbles	1⁄6–2.5 inches
cobbles	2.5–10 inches
boulders	diameter greater than 10 inches

The percentage of gravel in a soil may vary considerably depending upon the origin of the gravel deposit. Gravelly soils may contain 35–75% by volume of gravel; gravelly sands may contain 5–35% gravel, with the ratio of sand to clay of at least 9 to 1; gravelly clay may contain 5–35% gravel, and clay predominates over sand.

Gravel deposits resulting from continental glaciation yield a great variety of rock types because the long-traveled glacier has moved over and assimilated a great variety of rock. The predominant ones are the more resistant, such as quartzites, gneisses, and granites. Gravel deposits resulting solely from stream action are also composed of resistant rocks, but their makeup is less diversified since stream courses are shorter and, if flowing over horizontal sedimentary rocks, may consist of one or a few rock types. Some gravels possess a great range of stone size; others are quite uniformly granule, pebble, cobble, or boulder gravels. The last three, when consolidated, are termed conglomerates.

Gravels are used as aggregates in the making of concrete, as fill where great porosity is required for encircling septic tanks and leach fields, to hold slopes from sliding, and as remedial fill to retard erosion along railroad beds, highways, streams, and coastlines. Cobble and boulder gravel, together with huge blocks of angular stone, are referred to as "riprap" when used to thwart extreme erosion. Granule-gravel is used on floors of aquaria, reptile tanks, and birdcages.

Sand, like gravel, is an unconsolidated sedimentary deposit and a textural term for particles less than 1⁄12 inch. Finer material, still coarser than clay, is silt, ranging from 1⁄256–1⁄16 mm (4–62 microns). Frequently, sand is given a mineralogical connotation when having a makeup entirely or predominantly of quartz grains. Strictly speaking, however, sands and their consolidation products—sandstones—may consist exclusively or chiefly of other minerals, such as: calcite (calcium carbonate) in coral-fragmental sands or sand-textured limestones (calcarenites); feldspars (arkoses); gypsum (calcium sulfate), as in White Sands National Monument in New Mexico; basaltic black sands (Hawaii); olivine green sands (Hawaii); garnet; or magnetite. Sand (and sandstone) may be well-rounded, semi-rounded, or angular, depending upon the distance of grain transportation; the greater the transport distance, the greater the degree of rounding. Sands may originate in differing marine environments, from beach to deep trench, and in non-marine lake (lacustrine) or river habitats. And sands may accumulate on land (terrestrial) in a multitude of shapes and sizes as a result of glacial or wind activity (for example, in eskers, kames,or dunes).

Quartz (silica) sand is an aggregate in the manufacture of concrete. When applied to icy roads, silica sand (combined with an ice-melting component such as salt) provides improved traction. Certain molding sands are employed for casting purposes in industry. By volume, however, the principal use of silica sand is in the manufacture of glass.

To function cost-efficiently, a potential gravel and sand site must fulfill the following requirements:

1. ample deposit, particularly above the water table where pumping is unnecessary
2. quality of the gravel, so as to insure durability
3. availability of highways capable of handling heavy trucks

Gravels with considerable admixed shale are to be avoided, as they abrade easily and break up during periods of alternating freezing and thawing, thus undermining their compactness. To satisfy use for concrete aggregate, gravels must not react chemically with cements, must resist abrasion, and must maintain their cohesiveness during periods of freezing and thawing. In order to insure soundness, severely weathered stones, decayed rock, and decomposed soil should be removed prior to gravel introduction into concrete aggregate.

Gravel and sand pits are very transient features because their workable resources are normally quite limited. Abandoned pits quickly become overgrown with vegetation or are landscaped to conform to the adjacent topography by virtue of zoning regulations.

In the twenty-first century, a prosperous forecast awaits gravel and sand developers in the Hudson Valley. Gravel and sand have been, and will continue to be, dredged from the Hudson River in order to maintain a specified deep channel for oceangoing ships navigating the Hudson River to Albany. This excavated river sediment, marginal to the river, is a potential resource for human use. As the residential population grows, and with the continual immigration from metropolitan New York City, requirements for constructional materials will increase markedly. Inevitably, the gravel and sand (and cement) industries should enjoy expansion. With its navigable river, modern highways, and improved rail trackage and equipment, the Hudson Valley corridor will provide relatively affordable transportation to these and other geological resources. How rapidly this growth will materialize is linked to engineering and technological progress coupled with additional population demands. Accordingly, inventories of our natural resources are mandatory if we are to use them prior to "carpeting" by residential developments and business establishments. Once this blanketing has occurred, valuable unrenewable geologic resources will have been concealed forever.

Manganese

The existence of manganese-bearing minerals in Columbia County was first reported by Lewis C. Beck (1842) and W. W. Mather (1842) and was re-investigated by Nelson C. Dale (1919). The potential deposits occur in small swamps (often forested) and marshes (often treeless) usually between the elevations of 1,000–1,400 feet. Past occurrences have been reported in the Canaan, State Line, Hillsdale, Egremont, Copake, and Ancram Quadrangles, but they undoubtedly exist elsewhere in the county. These manganese oxide-bearing bogs occupy small depressions in the inter-hill saddles as divides, or in terrace-like benches at the foot of hills where the bogs serve as catchment basins for local drainage and as sources for small streams.

Bog manganese oxide, colloquially termed wad or black ochre, is a variety of the mineral psilomelane. This mineral is a soft, amorphous, brownish-black material, sometimes coal-like or sub-metallic, but usually showing an earthy luster. Other oxides or hydroxides of manganese may be intermixed. Common impurities are iron oxide, silica, and phosphorus and barium compounds.

Three growth habits of psilomelane are known in Columbia County. In decreasing abundance they are: nodules, clusters, and cement. Elliptical to spherical nodules, up to grape-size, within a kidney-looking exterior occur in light grey clay. In nearly dry bogs, the nodules are usually at the swamp outlets or in the source areas of the streambeds draining them. Less common are peppercorn-size spheroid clusters in a gray clay matrix. This habit appears to coincide with the horizon of the water table. The rarest is the cement habit, and the least manganese-pure is a glacial clay deposit with included boulders, cobbles, and angular rocks having occasional coatings of manganese oxide and interstitial films of manganese oxide. Because manganese nodules are plentiful on ocean and lake floors, modern technology will soon develop equipment that can harvest them, making extremely low-grade deposits on land economically unfeasible.

Peat

Unconsolidated deposits of semi-carbonized plant remains in a water-saturated environment, such as a bog or swamp, is termed "peat." It is an early stage in the development of coal, as the carbon content is about 60% and oxygen constitutes about 30%. Dried peat burns freely. Mixed with soils, it enhances the nutritional value and retains water for more luxuriant plant growth. Used as mulch, peat reduces surface water evaporation, insulates the soil from excessive heat, and inhibits weed growth. Ironically, the removal of peat from manganese-bearing ponds, swamps, or bogs is more worthwhile than the removal of the manganese. With future population growth within the county and, therefore an increase in gardening, Hudson Valley peat deposits may be highly sought-after.

Epilogue

In concluding this work, it seems appropriate to answer the many versions of the frequently asked question, "What led you into geology?"

My boyhood was spent during the Great Depression and World War II—a time before computers and television when more leisure activities centered on being outdoors. I was fortunate to live near the Buffalo Museum of Science, a fascinating educational resource. My parents were curious about nature and faraway places, but national and foreign travel was not possible for my family. The museum provided vicarious travel by offering free lecture programs by professional photographers, world travelers, and scientists. By accompanying my parents to these Sunday afternoon and Thursday evening presentations, I increased my knowledge of natural history and geography. This inspired me to frequent the museum to quiz the staff astronomer, biologists, and geologists with questions about their specialties. In retrospect, I was probably quite a pest! Today I have the opportunity to repay their kindness by fielding a variety of questions from interested visitors to my rock shop.

The museum also offered free weekend camping sessions with a focus on natural history education at Allegheny State Park in Cattaraugus County. Naturally I was quick to seize this opportunity. In addition to learning the customary camping routines—an adventure then seldom enjoyed except by the wealthy—I learned about rocks, trees and other plants, insects, frogs, snakes, birds, deer, and other creatures. What an exciting and varied world was revealed!

As a high school student I had the good fortune to have excellent teachers, especially in the sciences and history. Despite lacking the variety of media aids that teachers possess today, my teachers' stimulating methods and after-school enrichment sessions crystallized my interest in the sciences.

In my freshmen year at the University of Buffalo, I began my studies with the intention of majoring in chemistry. The world of plastics was in its infancy and suggested a promising profession. But would I be content with a life in a laboratory?

Sometimes a chance event has a profound effect on one's future. Such was the case for me during my sophomore year when I enrolled in my first geology course, taught by Dr. Reginald H. Pegrum. This extraordinary professor was the most outstanding teacher I had ever experienced. He had an uncanny gift for creating vivid pictures in students' minds and could make geologic processes seem to come "alive." When he described the formation, movement, and retreat of glaciers, I felt as if I was a spectator witnessing the events firsthand. When a ringing bell announced that his class had ended, he would smilingly say in a clear and forceful voice, "Ladies and gentlemen, class is over!" Slowly, in a trance-like state, we would depart to our next class. Dr. Pegrum demonstrated the mark of a great teacher. As a result of his influence, I became a geology major, and this subject became my life-long passion.

In addition to study in the classroom, a geologist must master field observation and collecting skills. Dr. Pegrum was superbly adept at explaining field techniques, structures, rock types, and mineral collecting during field trips. Advice on collecting fossil specimens was provided by an amateur fossil collector who happened to be the father of a student colleague of mine. Together, this father-son team seemed to know the best streams in western New York for obtaining excellent Devonian brachiopods, corals, trilobites, and other invertebrates, and they taught me the proper way to extract and label specimens.

As I reflect on my career, numerous factors stimulated and enabled me to enjoy the wonders of geology. The key factors, however, were curious and supportive parents, access to a wonderful science museum and staff, and inspiring and vibrant teachers.

Appendix I:

Where and How to Collect And Study Geological Specimens

"Go my sons, buy stout shoes, climb the mountains, search the valleys, the deserts, the sea shores, and the deep recesses of the earth. Look for the various kinds of minerals; note their characters and mark their origin."

Severinus (quoted by J. G. Wallerius, *Systema Mineralogium,* Preface, Vienna, 1778)

Necessary attributes and equipment of the collector and student of fossils, minerals, and rocks are an intrinsic curiosity, a pair of sharp eyes, steadfast perseverance, controlled patience, a few tools, and a zeal for learning more about the subject. Locating fossils is 90% luck and 10% skill; collecting and preparing them is 10% luck and 90% skill. Fossil extraction is a talent that blends paleontological knowledge, state-of-the-art technology, and great patience. Locating minerals requires more knowledge of their probable whereabouts and recognition of their likely host rock. Similar to fossils but more so, collecting and preparing minerals demands extreme care and diligence. On the other hand, rocks are the easiest to collect because they are more prevalent and not subject to critical damage when gathered. Rock samples should be fist- to golf ball-size and showing a fresh fracture so that the mineral content can be readily determined. Stones can be heavy, so be selective in quality; leave some for the next visitor. Labeling should be done in the field. Notation should include location of find, date, name of collector, formation name (if known), and the name of the fossil, mineral, or rock (if known).

Fig. 129. Obtain permission before collecting on private property. Don't block driveways or leave garbage! Drawing by Helen Fischer. Printed with permission of the New York State Museum, Albany, NY.

Potential collecting sites are field outcroppings, streambeds and banks, beaches and rocky coasts, road and railroad cuts, gravel and sand pits, stone quarries (preferably inactive), and temporary excavations. Unless otherwise prohibited (such as in parks, along parkways or interstate highways, or in cemeteries), collecting without prior approval is permissible along federal, state, county, and town roads as these right-of-ways (usually between the utility poles and the pavement) are in the public domain. Certainly, attention should be addressed to traffic volume and the parking of your vehicle completely off the roadway. Obviously, before attempting to collect on private property, permission should be obtained from the landowner. After securing such right to trespass, common sense dictates that you should park your vehicle in an innocuous place, ignore the owner's animals, keep your children under control, leave the owner's possessions alone, close fence gates, and take everything with you that you came with. Thank the owner verbally or with specimens or a gift—you may wish to return!

Fig. 130. Leave all gates as you found them!
Drawing by Helen Fischer. Printed with permission of the New York State Museum, Albany, NY.

Fig. 131. Don't collect only for the dollar signs!
Drawing by Helen Fischer. Printed with permission of the New York State Museum, Albany, NY.

Occasionally, fossils and minerals may be found loose, having weathered out of their enclosing sediment or rock. More often, however, they must be removed by mechanical means. A variety of excellent rock-breaking tools manufactured by Estwing Manufacturing Company in Rockford, Illinois, are available from rock shops, and in some cases are available in well-equipped hardware stores. Pick- or chisel-end hammers and one- or two-handed sledges offer a good selection along with various types of nylon-sheathed cold chisels (Fig. 136). These tools have been tempered for rock breaking; carpenter and machinist's hammers are not. A small magnifying (10 ×) loupe or hand lens is a must since very small fossils and crystals may elude your naked eye. Wearing your magnifier on a cord or chain around your neck reduces risk of loss and frees your hands for other uses. Depending on the duration of your collecting trip and the distance to be traveled (walked?) to the collection site, other items you may need are food and water, compass, topographic map, camera, pencils, marking pens, canvas drawstring bags, notebook, labels, and wrapping material (bubble wrap, paper toweling, cotton). Absolute necessities for the serious collector are sturdy, over-the-ankle walking shoes, safety goggles, insect repellent, and a many-pocketed jacket. Durable cowhide or deerskin gloves will prevent hand chafing and sliver-like rock fragments from becoming imbedded in your hands.

Specimen collecting in sediments or semiconsolidated strata is very different. Hammers and chisels are rarely employed. Instead, shovels, trowels, knives, nutpicks, and whiskbrooms are used. Cobbles or pebbles in gravel will require breaking in order to display a fresh fracture for mineral identification. In general, loose minerals need only be water-washed or cleaned with compressed air. Fossils (usually bones, teeth, and shells) receive similar cleaning techniques and sometimes a mild vinegar wash. For fragile specimens, brushing with shellac or liquid plastic will inhibit breakup. For long-haul transport and eventual storage, bones may be bandaged with burlap or foil, then soaked in plaster of Paris thereby insuring rigidity.

At home or in the laboratory, excess rock matrix may be discarded using a trimming hammer, professional hydraulic rock trimmer, or rock saw (with a tungsten carbide or diamond blade). This operation must be performed with great care so as not to destroy the mineral or fossil. Should this near-disaster happen now, or during field collecting, epoxy or a similar cement is an indispensable repair material. The removal of obscuring matrix can be accomplished with teasing needles or nutpicks under magnification, but this is an arduous, time-consuming task. Quicker results can be obtained with inexpensive small power tools such as the Dremel or Vibrograver brands. Detergents and chemical reagents are frequently employed for cleaning minerals—however,

Fig. 132. The "well-equipped" fossil/mineral collector. Printed with permission of the New York State Museum, Albany, NY.

Fig. 133. Pay attention to what's going on behind you! Printed with permission of the New York State Museum, Albany, NY.

caution should be exercised when working with acids and alkalis; protective gloves are a must and inhalation of fumes should be avoided. Siliceous fossils may be isolated from limestone (calcium carbonate) by immersion in dilute hydrochloric (muriatic) acid, which frees the unaffected fossils from the soluble matrix. Similarly, phosphatic fossils (calcium phosphate) may be freed from limestone by immersion in acetic or formic acids, which slowly dissolve the soluble limestone. In recent years, museums, geological surveys, and university geology departments have improved fossil and mineral enhancement by blasting with compressed air charged with mild abrasive such as sodium bicarbonate or dolomite. This type of "airbrushing" reveals exquisite detail, previously not readily obtainable, and removes unwanted matrix within a relatively short time.

Additional techniques are imperative for those who wish to study and identify fossils whose classification is based upon interior structure (protozoans, sponges, corals, bryozoans, cephalopods, algae, and petrified wood). To accomplish this, thin sections and polished surfaces must be made. Specimens can be sliced with a tungsten-carbide or diamond-saw. For those who possess neither, the same result may be accomplished by pressing the specimen against a power-driven carborundum or emery grinding wheel. The same result may be obtained by substituting a flat pane of plate glass on which a water-sludge of progressively finer grades of carborundum powder has been applied. Using a circular hand motion of the specimen will produce a flat matte surface. Polishing is done on a separate glass plate (covered with cloth to avoid contamination from coarser abrasives), upon a water mixture of aluminum oxide, tin oxide, cerium oxide, or zirconium oxide. Iron oxide (jeweler's rouge) was widely used in the past, but is slow and stains clothing and skin red. As before, hand polishing in a circular motion is recommended. An effective way to obtain a polished look without going through all the aforementioned procedure is to coat the cut-section with clear plastic spray (Krylon) or brush thinly with clear shellac. Prior to polishing, it may be instructive to have a transparent peel made from a sawed section of a fossil. This is accomplished by slightly etching the ground surface (it must be limestone) with dilute hydrochloric (muriatic) acid, washing with water, and flowing on a layer of collodion or parlodion; these acetate peels, when hardened and lifted off, record the cellular structure of the fossil. Thin sections of a rock, mineral, or fossil may be prepared by gluing a one inch square, $\frac{1}{8}$″–$\frac{3}{16}$″ (3 mm–5 mm) thick-sawed slab to a glass slide with Canada Balsam, bioplastic, or similar adhesive. Grinding the slab down to 0.2–0.4 mm will reveal microscopic details. Though some minerals are opaque, many minerals and most fossils become translucent or transparent by this operation. Naturally, one must have access

Fig. 134. I forgot my steel-toed shoes! Printed with permission of the New York State Museum, Albany, NY.

Fig. 135. I should have worn my safety goggles! Printed with permission of the New York State Museum, Albany, NY.

to a microscope to observe and study optical properties and micro-makeup. Where fossils occur only as molds, it may be useful to make a cast so as to facilitate identification. Molds should be greased with oil so that the cast material will not stick to the mold. Casts may be composed of modeling clay, dental wax, plaster of Paris, or liquid rubber. The rubber is applied by brush in layers; it is less messy but ammonia fumes from the solvent may be irritating.

Photographing minerals and fossils is especially rewarding and an efficient way to document your collection. When in the field, it may prove impossible to remove choice minerals, fossils, or specific structures. A field photo will record the occurrence. In your pictures, make certain to position a common object of known size to establish scale. Hammers, pencils, a coin, or a lens cap are commonly used. In the lab or at home, particular attention should be given to the choice of lights and positioning of them with a complimentary background during picture taking. With minerals, particularly transparent or translucent ones, sidelighting or backlighting is frequently preferred. Because fossils are opaque, low relief ones should receive low incident light and high relief ones should receive high incident light. "Smoking" with a thin haze of heated ammonium chloride crystals or fumes of burning magnesium ribbon may enhance highlights on delicately ornate fossils (trilobites, ostracodes, crinoids, and other echinoderms). Jeffrey Scovil's excellent book, *Photographing Minerals, Fossils, & Lapidary Materials,* is a must for the serious photographer.

Resist the temptation to dump unlabelled and unwrapped specimens in a container that will disappear into the attic or garage. Your specimens should be treated as natural objects of beauty, perfection, and with a history to be revealed. As such, the diligent collector should, like the consummate scientist or efficient file clerk, maintain his or her material in an orderly fashion—properly labeled in as dust-free a location as possible. Showcases, drawers, plastic transparent boxes are all adaptable to small- or medium-sized specimens, which may be organized by rock, mineral, or fossil type, or geographically. A well-lit, uncluttered display of your specimens will excite the interest of your guests and add elegance to your home.

Identifying your material need not be a problem. There are many splendid, well-illustrated books available which will educate you further concerning rocks, minerals, fossils, and geology in general. You should have a select number in your personal library. Professional and amateur mineralogists, paleontologists, and earth science teachers will assist in identifying your material personally or direct you to references that can provide answers to your questions. Beware of alleged geologic information on the Internet; unless reputable geologists have furnished it, it is suspect.

Searching, extracting, identifying, photographing, and curating are time- and effort-consuming, but the rewards are great if you've made a choice "catch." Of course there will be days when you come home unsuccessful in your quest. But the memory of such days is quickly erased when you uncover a worthwhile specimen. Remember, after splitting a rock open, you are the first human to gaze upon this mineral or fossil from antiquity. Happy hunting!

Fig. 136. Some basic tools of the fossil/mineral collector. Drawing by Wayne Trimm. Printed with permission of the New York State Museum, Albany, NY.

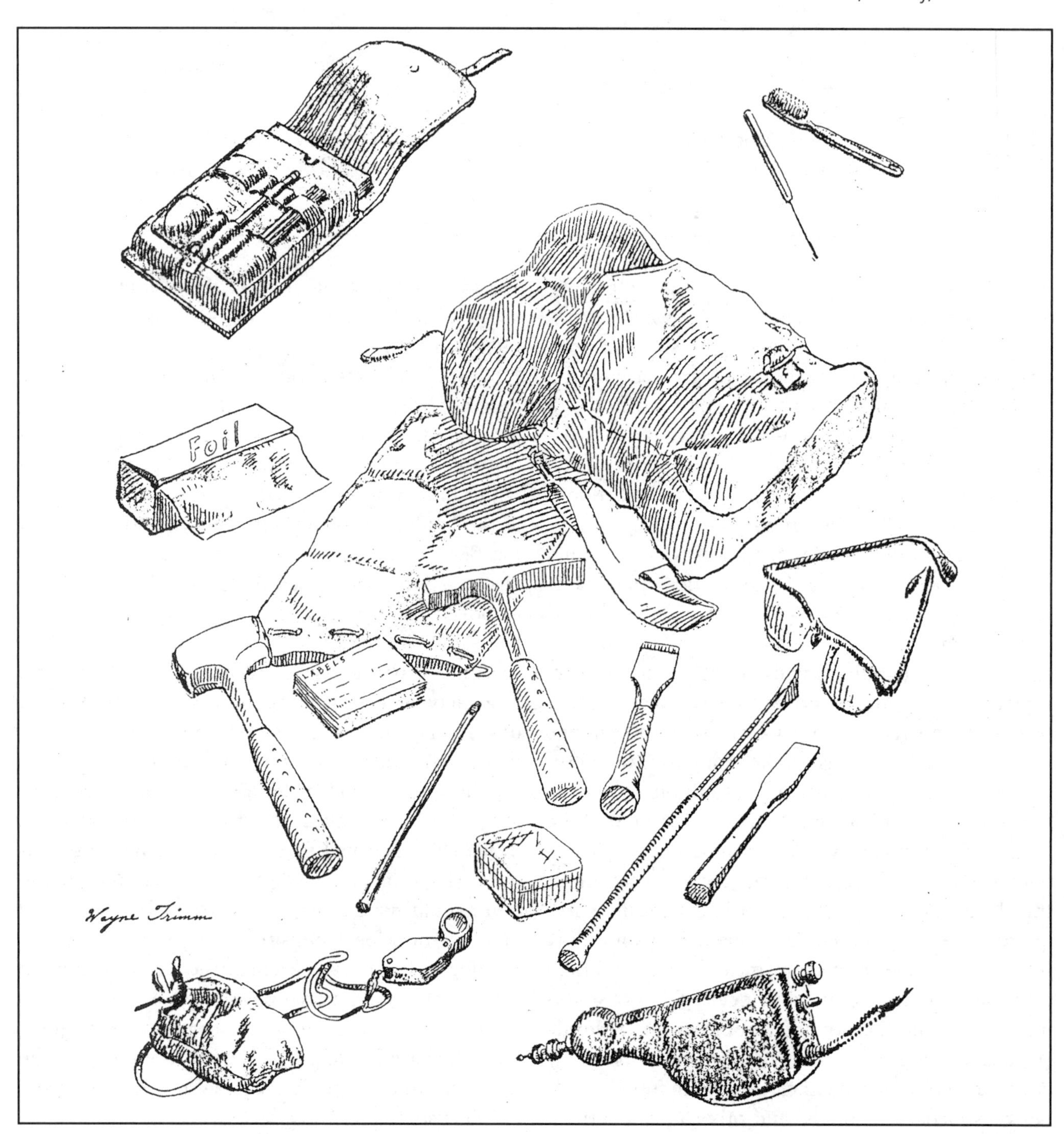

Appendix II:
Historical Synopsis

"A science which ... forget[s] its founders is lost."
Alfred North Whitehead

Columbia County's Pioneer Geologist: Amos Eaton

Historians of geology naturally focus on the infancy of the science in North America—the first four decades of the nineteenth century. Within this time frame, the accomplishments of Columbia County's Amos Eaton are paramount. Born in New Concord, Chatham Township, on May 17, 1776, Eaton epitomizes the budding of the natural sciences in an emergent nation. As a talented youth, he possessed a broad spectrum of interests including oratory, blacksmithing, and surveying. From 1791–1795 Amos was a student at the now defunct Spencertown Academy (today the restored building is a popular cultural center for area residents). After receiving his B.A. degree from Williams College in 1799, he served as a law apprentice with Elisha Williams at Spencertown and continued his law studies in New York City. Here, Eaton's destiny became "fixed" under the tutelage of Columbia College physicians David Hosack and Samuel Mitchill—the premier natural scientists in colonial America.

Nevertheless, the next few years were to be an odd interlude in Eaton's pursuit of a scientific career. This difficult period began when his first wife, Mary (Polly) Thomas, died on September 15, 1802, at the age of 21—leaving Eaton destitute, without a job, and with an infant son. With his law training as his only viable credentials, Eaton gained a measure of financial solvency by becoming a land agent, rent collector, attorney, and all-around supervisor for the land-wealthy John Livingston—specifically for his estate in Greene County.

Eaton's now-comfortable lifestyle was shockingly interrupted on January 7, 1811, when he was indicted by the grand jury of Greene County for forgery allegedly related to land speculation. On August 16, 1811, he was convicted and sentenced to life at hard labor—a most severe penalty for such an offense. Eaton was incarcerated in Newgate Prison in Greenwich Village, Manhattan. Imprisonment coupled with suffering the deaths of his parents, his beloved second wife Sally Cady, and son Charles Linnaeus would for most bring on extreme melancholia, but Amos Eaton was made of very stern stuff indeed.

While in prison, Eaton was befriended by and learned botany from the warden's son, John Torrey, who later distinguished himself as a famous botanist. During a prison inspection trip, New York City Mayor DeWitt Clinton, also an amateur botanist, detected great scientific potential in Eaton. Believing in his innocence, Clinton persuaded Governor Daniel D. Tompkins, later vice president under James Monroe, to issue a conditional pardon; it was finalized November 17, 1815. Earlier that year, Williams College had recognized Eaton's botanical expertise and bestowed upon him an honorary M.A. degree. Freed from Newgate, Eaton matriculated at Yale University in 1816 to enhance his botanical knowledge under Professor Eli Ives and to learn geology and mineralogy from the celebrated Professor Benjamin Silliman.

His quest for a life of science continued in 1817 when he returned to Williams College as instructor in botany, geology, and zoology. There, Eaton authored his popular *Manual of Botany* (1817), which sold for $1.75, went through eight editions and contained descriptions of 5,267 species of plants—an immense and influential contribution.

Now, sporting three scientific "hats," Eaton traveled to colleges, prep schools, and village squares as an itinerant lecturer in natural sciences. At the invitation of the then-Governor DeWitt Clinton, Eaton lectured to the New York State Legislature. The subject was Clinton's "Big Ditch"—his pet project for construction of a canal that would be an economic boost for towns along its route. Clinton's strategy paid off as Eaton's lively and informative style was so charismatic that the staid politicians appropriated funds to support the gathering of information about the state's mineral and economic resources prior to, and during, construction of the Erie Canal. Also in 1818, Eaton authored America's first geological textbook, *An Index to the Geology of the United States,* wherein he included a rock- and time-classification scheme, a local guide to field trips, and a geologic cross section from the Catskill Mountains to the Atlantic Ocean.

Through 1824, Eaton lectured to the populace on matters of natural history like a wandering evangelist, amassing over 2,000 public appearances. He was so charismatic that he had audience "converts" who acted as disciples. This outbreak of scientific oration came to the attention of patroon Stephen Van Rensselaer, who was prudently assembling outstanding faculty for his newly established Rensselaer Institute (renamed the Rensselaer Polytechnic Institute on April 18, 1861) in Troy, New York. In November 1824, Amos Eaton was appointed first Senior Professor of Geology—a department that was to enjoy a productive and memorable legacy for over 175 years.

As a comprehensive science professor, Eaton taught chemistry, geology, mineralogy, physics, surveying, and civil engineering. In the academic title field, he introduced the degrees of Bachelor of Natural Sciences and Bachelor of Civil Engineering. For his era, Amos Eaton would prove to be a pedagogical maverick who served as a catalyst for innovative and unconventional teaching techniques. His "Rensselaerian Plan" included mandatory "learning-by-doing"; students were required to participate in field collecting, laboratory experiments, and practice teaching. This was a bold departure from the traditional method of that day, in which professors formally read from prepared notes, and called on students to recite from memory. His chemistry students were required to perform approximately 500 experiments and his geology students were to collect and describe over 100 rocks and minerals; each science student was required to give 30 lectures. Eaton had never been a "closet" naturalist but was always a strong proponent of field observation and collecting; witness to this is his lifetime compilation of 17,000 miles of field trip work! Perhaps even more unorthodox was Eaton's open encouragement of women students who possessed a scientific bent—a most radical opinion for that period! Among these were his daughter Sarah Cady Eaton, Laura Johnson (his fourth wife's sister), Almira Lincoln Phelps (sister of Emma Willard, the founder of the famous school for girls), and the emancipator Mary Lyon.

Amos Eaton attained the rare achievement of becoming a colossus in two sciences—geology and botany. Following publication of his geology index and botany textbook he gained further geologic fame for his work entitled *A Geological and Agricultural Survey of the District Adjoining the Erie Canal* (1824) under the auspices of his perennial benefactor DeWitt Clinton. Also, while at Rensselaer, he wrote a *Geological Textbook* (1830), which sold for $1.50 and included the first colored geologic map of New York State (Fig. 137). In related fields, he was teacher to the inventor of the Thatcher Slide Rule and the designer of the Ferris Wheel. In addition, Eaton taught three quarters of the famous bridge builders of the nineteenth century—including John Roebling, architect of the Brooklyn Bridge. Eaton's contributions to American geology included being mentor to an entire generation of pioneer geologists, many of who comprised the workforce for the rash of newly established State Geological Surveys, twenty-five of which originated from 1830 to 1858.

As a passionate peddler of science, Amos Eaton provided an impetus to the natural sciences, particularly geology, at a time when scientific specimens had been little more than curiosities of nature. At the time of his death, he was the principal sage of American geology. Eaton died on May 10, 1842, one week short of his sixty-sixth birthday, in Troy, New York, and is buried there in Oakwood Cemetery. For more on Eaton's life, see McAllister (1941) and Fisher (1978).

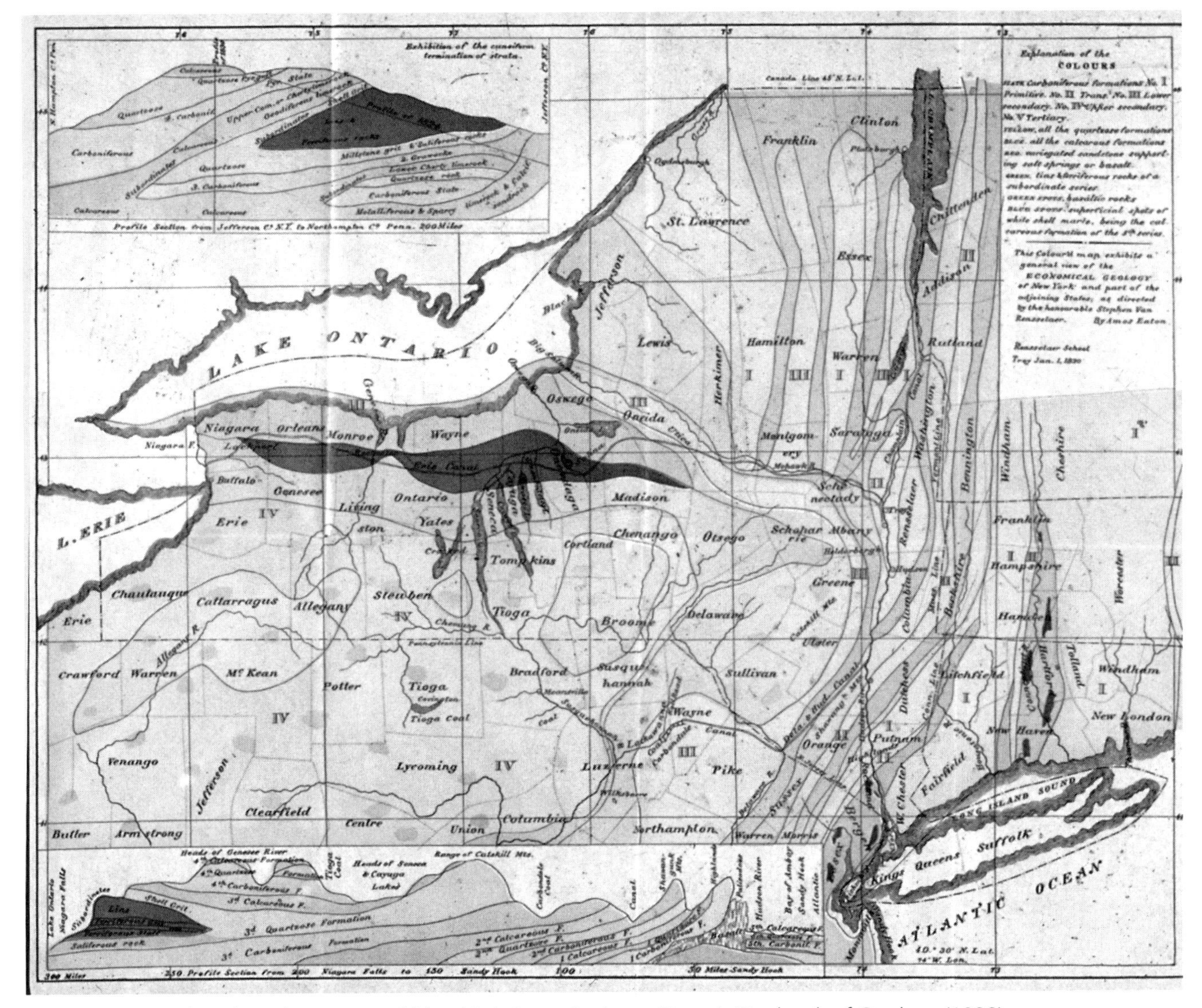

Fig. 137. First colored geologic map of New York State. In Amos Eaton's *Textbook of Geology* (1830).

Politics and the New York State Geological Survey

April 15, 1836, was a turning point in geologic undertaking in New York State. On that momentous day, Governor William Marcy, an ardent proponent of conservation, signed into law the establishment of the Natural History Survey of New York, which included the New York State Geological Survey. This was the brainchild of the astute New York Secretary of State John Adams Dix. Were it not for his cogent arguments before a skeptical legislature stressing the value of a scientific and economic resources assessment of the state, the survey would not have been undertaken. Dix did his homework well. The climate was utopian as far as the social atmosphere and political prudence were concerned. The relatively unexplored youthful country was recovering from the depression-caused War of 1812 and was eager, energetic, and longing to expand "new frontiers" in science and the humanities. Fortuitously, there existed a clique of progressive politicians who were instrumental in catalyzing these desires. In addition to Dix, there were DeWitt Clinton, William Marcy, and William H. Seward, governors of New York, and Martin Van Buren, president of the United States and vice president under Andrew Jackson. As a result, the New York legislature appropriated $104,000 for a statewide survey—undertaken on foot, on horseback, and by wagon—to be accomplished over a four-year period. In 1840, Governor William H.

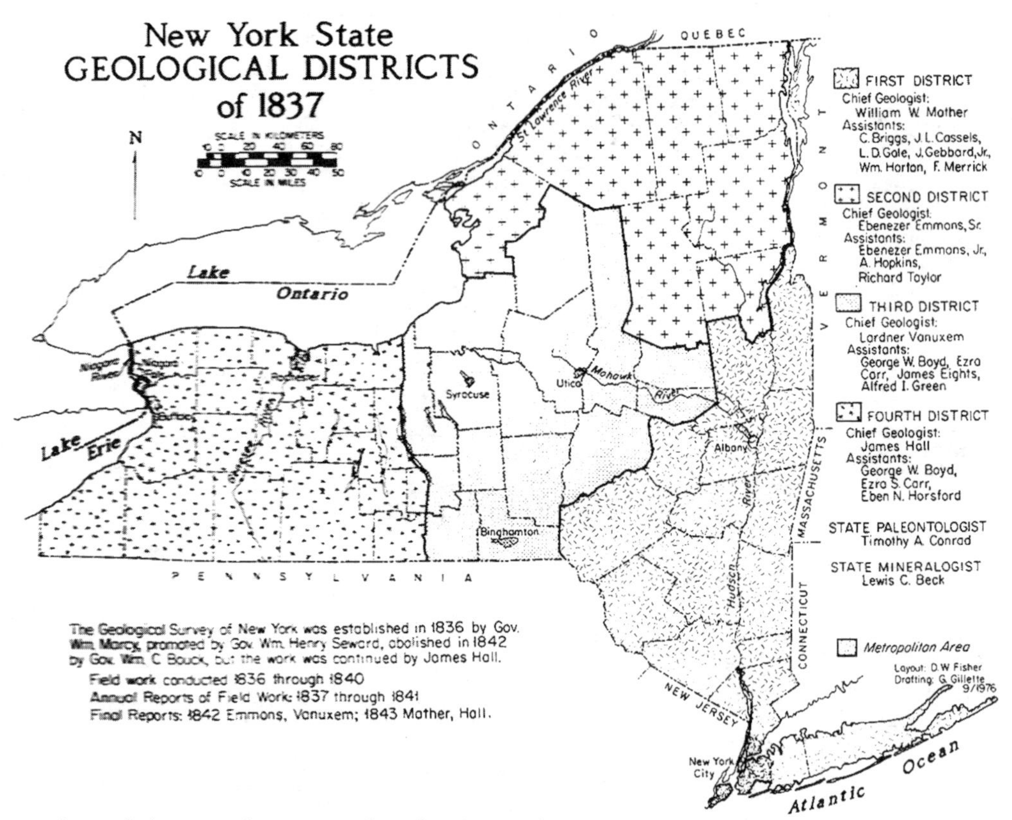

Fig. 138. New York State geological districts of 1837. Drafting by Gwyneth Gillette. Printed with permission of the New York State Museum, Albany, NY.

Seward, later famous for the purchase of Alaska while he was United States Secretary of State, extended the survey to five years. These investigations and their resultant publications did more to promote geology in nineteenth-century America than any other of a similar scope.

On the advice of Amos Eaton and Professor Edward Hitchcock of Amherst College, the Geological Survey was apportioned into four districts (Fig. 138). Both of these respected geologists were in failing health and were considered too old for strenuous fieldwork. Nevertheless, Eaton's sage advice was constantly sought after, especially in the hiring of staff. A case in point was his succinct recommendation for Caleb Briggs, Jr.: "Mr. Briggs is amiable, and industrious, and poor. This tells all I can tell."[20] (Regrettably, Eaton did not live to see the final comprehensive District Reports of the Survey whose seed he had nurtured, for he died on May 10, 1842.)

A principal State Geologist with equal authority and, it was hoped, equal competency headed each district. William Williams Mather (Fig. 139), a retired army officer and teacher of geology and chemistry at the U. S. Military Academy at West Point and formerly at Wesleyan University, was in charge of the First, or Eastern District. Ebenezer Emmons, Sr., chemist, mineralogist, doctor of medicine, and professor of chemistry at Williams College, headed the Second, or Northern District. Lardner Vanuxem, trained at the École des Mines in Paris, France, was chosen to head the Fourth, or Western District, but after one year was transferred to the Third, or Central District. In the second year of the survey, twenty-five-year-old James Hall succeeded Vanuxem as chief of the Fourth District. Each principal investigator had from two to eight proficient assistant geologists at differing times (Figs. 140, 141, 142). Published annual reports and four final district tomes were contractual stipulations, as were summary tomes on New York's minerals and fossils. Only the annual reports, the four summary district reports, and the mineralogy volume by Lewis C. Beck (1842), physician and professor of chemistry, mineralogy, botany, and zoology at Albany Medical College and Rutgers University, were completed.

Timothy A. Conrad, a conchologist (a zoologist who specializes in shells), was specifically hired to research the fossils of New York State because the other principal scientists were not trained in paleontology. Conrad was assigned responsibility for the Third District in 1836 since it was known from amateur collectors that this area teemed with fossils. Nevertheless, Conrad was none too fond of fieldwork and, as progress lagged, the Third and Fourth Districts were re-aligned; Vanuxem was transferred from the Fourth to the Third District and young James Hall (then assistant to Emmons in the Second District) was made chief of the Fourth District. In 1837, Conrad was given the title of "Paleontologist of the Survey," a desk job of studying, classifying, and describing the prodigious array of unidentified fossils amassed during the first year's fieldwork of the entire Survey. Still, his dilatory attitude prevailed. Undoubtedly, he was overwhelmed by the lavish number of fossils to be investigated or perhaps he had an aversion to writing. Although "married" to New York's Paleozoic fossils by contract, Conrad surreptitiously carried on "extra-marital affairs" with Tertiary fossils from New Jersey and Maryland! As the Atlantic Coastal Plain marine sediments hold a preponderance of molluscan shells, it is understandable that Conrad would return to his "first love." A volume incorporating locations, descriptions, and illustrations of the fossils collected never materialized.

When Governor William C. Bouck terminated funding for the Survey in 1843, Conrad hastily departed without bothering to resign. Beck, however, continued research on mineralogy and medicine and achieved an international reputation for his work on pure foods and drugs. Mather

Fig. 139. Principal staff of the New York Geological Survey (1836–1841). Printed with permission of the New York State Museum, Albany, NY.

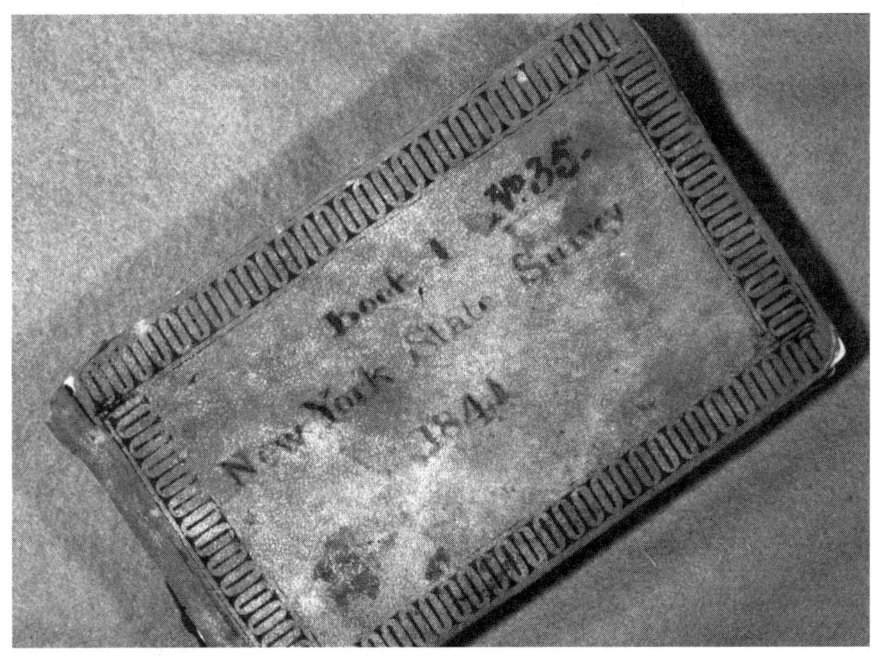

Fig. 140. Book 1, No. 35, New York State Survey 1841. Richard C. Taylor's field notebook. It may be assumed that Taylor was an assistant field geologist to Vanuxem in the Third District. This is the only known surviving field notebook of the 1836–1841 New York State Geological Survey. Archives of the New York State Library, Albany.

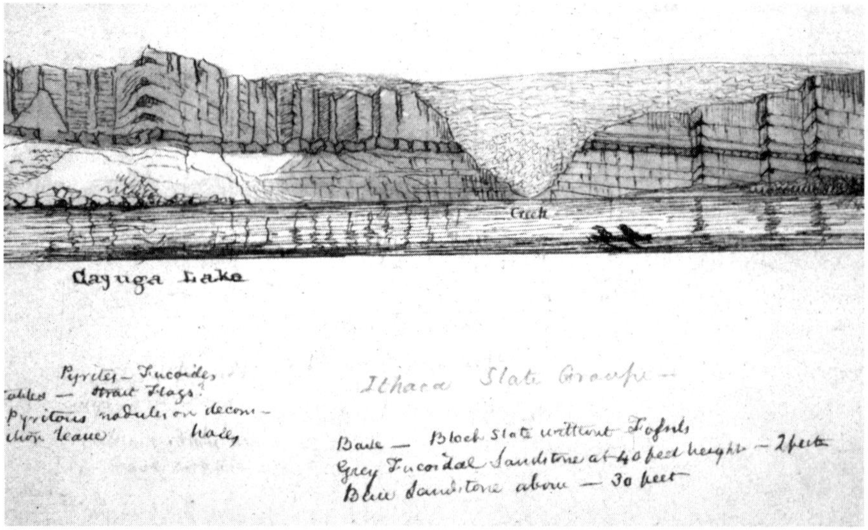

Fig. 141. Sketch with geologic notes by Richard C. Taylor of Ithaca "slate" along Cayuga Lake, 2½ miles southwest of Scott's Tavern. From the only known surviving field notebook of the 1836–1841 New York State Geological Survey. Archives of the New York State Library, Albany.

became active with the growing Ohio Geological Survey and became Professor at Ohio University in Athens, Ohio. Eventually he became acting president of that university. Vanuxem retired to his farm near Bristol, Pennsylvania, leaving Emmons and Hall to compete for funding from the legislature (Fig. 143) to complete the work to which Conrad had been assigned. In addition, funding was desired to continue tapping the fossil wealth still uncollected from New York State's sedimentary rocks.

Ebenezer Emmons, Sr., was an avid collector of rocks, minerals, plants, and insects—a typical avocation for aspiring naturalists of that period. In 1818 Emmons graduated from Williams College and thereafter enrolled in Albany Medical College. Subsequently, he practiced general medicine and surgery for 15 years in Berkshire County, Massachusetts, and from 1838–1850 held the unlikely dual professorship of Chemistry and Obstetrics at Albany Medical College. To the unnamed and almost unknown mountains of northern New York, Emmons applied the name *Adirondack* (1838) and named its highest peak, Mt. Marcy, to honor the governor who advocated state-funded scientific endeavors. More-

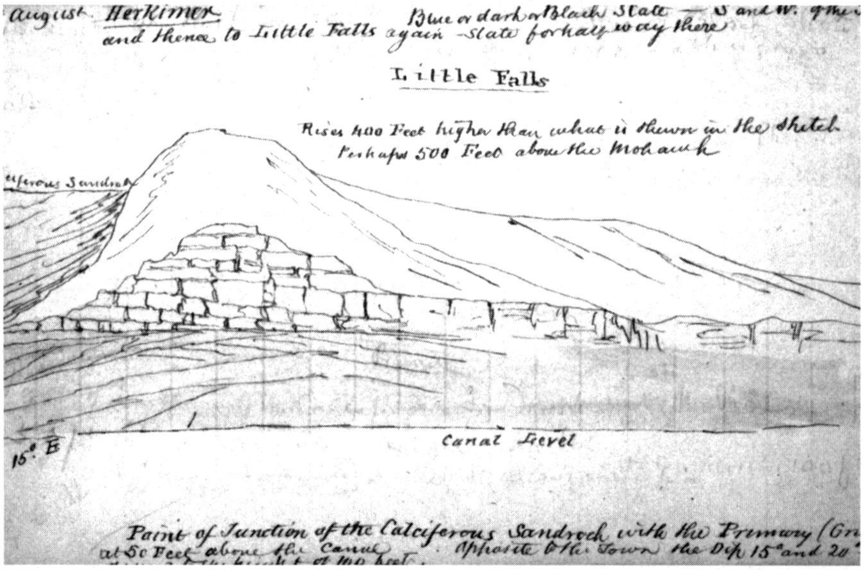

Fig. 142. Sketch by Richard C. Taylor of junction of the "Calciferous Sandrock" (Little Falls Dolostone) with underlying "Primary Gneiss" (Proterozoic charnockite) at Little Falls, Herkimer County. From the only known surviving field notebook of the 1836–1841 New York State Geological Survey. Archives of the New York State Library, Albany.

over, at Emmons' suggestion, the four principal geologists of the young Geological Survey named the fossiliferous sedimentary rocks the "New York System." It was (and is) a more complete stratigraphic standard than the Cambrian, Ordovician, Silurian, and Devonian Systems then being promulgated by Adam Sedgwick, Charles Lapworth, and Rodney Murchison in the highly deformed and stratigraphically incomplete sections in Great Britain. Nevertheless, the youthful status and scarcity of accredited geologists in North America willing to support the "New York System" did not prevail when compared with the patriarchal, renowned, and numerous European geologists who favored the convictions of British workers and their terminology. Consequently, the name "New York System" quickly fell into oblivion.

Governor Bouck appointed James Hall to the position of State Paleontologist in 1843 and Emmons as State Agriculturist. Making the most of this ostensibly demeaning position, Emmons produced four monographs on the *Agriculture of New York* (soils, analyses of grains, fruits and vegetables, and insects). Within the soils tome (1846, pp. 45–112), he craftily included his notions on the "Taconic System" of rocks in eastern New York. Emmons' affirmation of the greater age of Taconic rocks rested on the discovery in 1844 of the trilobites *Elliptocephala asaphoides* and *Atops trilineatus* (members of what was termed "Primordial Fauna") by his entomologist friend Asa Fitch (officially the first government entomologist). Later paleontological discoveries by Silas W. Ford[21] (1871–1880), an amateur from Troy, New York, and William Buck Dwight (1885), Professor of Geology at Vassar College, corroborated the existence of Early Cambrian strata (using a suite of "Primordial Fauna" fossils) in the Taconics. Despite this confirming evidence of a pre-Late Cambrian (Potsdam) age, James Hall was stubbornly resolute in his belief that no Taconic rocks could be older than Late Cambrian. Hall focused on other geographical areas in need of paleontological study,

Fig. 143. State Paleontologist James Hall demanding more money for research on fossils from the apathetic budget committee of the New York State Legislature. Drawing by Jennis Cortez. Printed with permission of the New York State Museum, Albany, NY.

often holding simultaneous State Geologist positions in several states (Fig. 144). But he did not forsake New York State's fossil wealth, as his most remarkable accomplishment is the publication of fourteen quarto volumes, *Paleontology of New York* (1847–1894)—the most unique paleontological achievement in nineteenth-century North America.

Apropos to any analysis of Taconic Geology was the comprehensive discussion by William W. Mather (1843), *Geology of New York, Part 1, Geology of the First Geological District*. This weighty reference consists of 653 text pages and 44 plates (many hand-colored) and was completed with the help of eight assistant field geologists. Mather's descriptions of rock and structure are scholarly for that period, but being unschooled in paleontology he could not justify support for a "Primordial Fauna."

Professor Chester Dewey of Williams College introduced the name "Taconick" in 1819. He also prepared the first geological cross section from Williamstown, Massachusetts, to Troy, New York, in which the intricacy of the Taconic Mountains was first suggested (1824). While a member of the Williams College faculty, he established the college's first chemistry course that included a generous dose of mineralogy. Dewey's favorite pupil (like his professorial "sidekick," Eaton), and also his field assistant, was Ebenezer Emmons, whom he extolled as "an indefatigable and acute observer." Later, when Emmons stood assailed as an incompetent during his conflict with James Hall (in the Taconic Controversy, to be detailed later), Dewey's esteem for Emmons was a public demonstration of support for Emmons' assertion of the antiquity of certain Taconic rocks. Although Chester Dewey is primarily remembered for his later work as a mathematician at the University of Rochester, his earlier career produced significant contributions to Taconic geology.

For additional information on the New York State Geological Survey, see Fisher (1978, 1981).

Fig. 144. James Hall's many "hats." Drawing by Jennis Cortez. Printed with permission of the New York State Museum, Albany, NY.

The Taconic Controversy

The Taconic Mountains have created a unique arena for geologic confrontation. Structural complexity and scarcity of easily identifiable fossils hamper efforts to determine precise stratigraphic arrangements and ages of the sometimes physically similar rock formations (mapping units). This drama of differing views has persisted for over 150 years as the Taconic Controversy.

It commenced in the mid-nineteenth century due to opposing correlation viewpoints; correlation is the process of demonstrating that geographically separated strata are equivalent, for the purpose of assembling a more complete rock "record" or geologic history. Index fossils are perhaps the most valuable tool in the geologist's "correlation toolbox." In spite of unambiguous (though relatively scarce) fossil evidence, steadfast disagreements regarding structural mechanics, ages, and relative sequences of strata continued to evolve throughout the twentieth century. Most participants handled it well, others not so well because personalities, pride, and prejudice took precedence over the earlier and subsequent accumulation of field evidence. Because facts tend to become obscured and theories fall short of their preconceptions, it is probably not so strange that the geologic history of the Taconic Mountains has resisted attempts at closure. Discord continues today, but on a greatly reduced and less antagonistic scale.

This dispute began during the mid-1840s between two charter members—by now familiar to the reader—of the New York State Geological Survey. Dr. Ebenezer Emmons, Sr., and Dr. James Hall were principal geologists of the Second and Fourth Geological Districts, respectively. Though originally close colleagues (Hall was a student of Emmons at the Rensselaer Institute and, later, field assistant in the Second District), their relationship degenerated and they became bitter adversaries (Fig. 145). The animosity of these two titans of geology festered, owing to divergent interpretations of the relative ages of Taconic rocks. Emmons was emphatic that his Taconic System, with its "Primordial Fossils," and all Taconic rocks were older than the Late Cambrian-aged Potsdam Sandstone—which up to this time had been accepted as the oldest Cambrian unit in New York State. The Potsdam Sandstone originated as an ancient beach sand and rests with a pronounced erosional break (termed an *angular unconformity*) on the much older (Medial Proterozoic) highly deformed metamorphic rocks of the Adirondack Mountains (Fig. 146). James Hall was pugnaciously resolute that all Taconic rocks were the same age as the undeformed strata west of the Hudson River, and thus considerably younger than the Potsdam Sandstone. The once amicable association of the protagonists immediately deteriorated to bitter animosity on Hall's part and to a more passive status of "agree to disagree" on Emmons' part.

Fig. 145. James Hall (left) versus Ebenezer Emmons, Sr. (right) on the Taconic Controversy. Drawing by Jennis Cortez. Printed with permission of the New York State Museum, Albany, NY.

Fig. 146. Angular unconformity at white line. Upper Cambrian Potsdam Sandstone on Middle Proterozoic Prospect Mountain Gneiss, along north side of NY 9L near southern end of Warner Bay, 5 miles northeast of Lake George Village.

It should be interjected here that during the latter half of the nineteenth century, Albany, New York, was the American Mecca of invertebrate paleontology (Fig. 147). Hopeful paleontologists-to-be, unable to pursue formal training at American colleges and universities, were eager to work with Hall and benefit from his expertise on fossils. Their tenure with Hall was often tumultuous; however, the irascible sage wielded enormous influence by virtue of his two statutory positions as State Paleontologist and State Geologist.

Aggravating this discordant quarrel between Emmons and Hall was the issuance of a geological chart in 1850 by James T. Foster, a schoolteacher in North Greenbush, Rensselaer County. This chart illustrated New York State rock formations and their relative ages and

Fig. 147. James Hall's School of Paleontology. This building is still standing in Lincoln Park, Albany. Note young men entering on right, and old men leaving on left. It takes years to become a professional paleontologist! Drawing by Jennis Cortez. Printed with permission of the New York State Museum, Albany, NY.

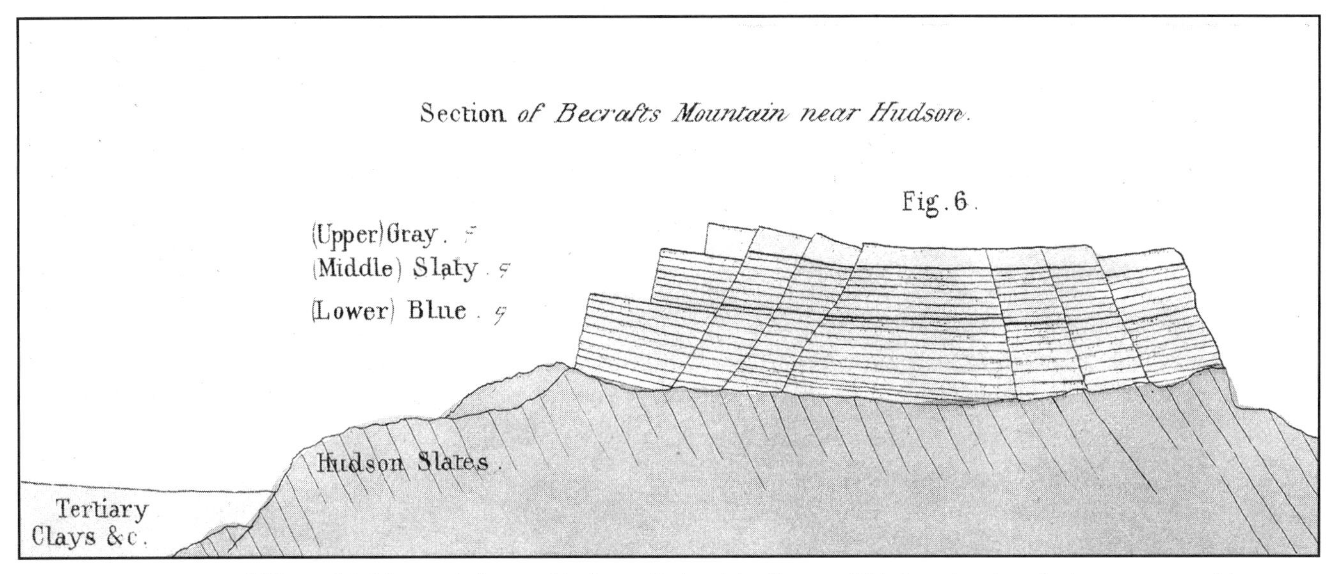

Fig. 148. A section of "Becraft's Mountain" near Hudson, Columbia County. This hand-colored plate appeared in Mather's *Geology of New York, Part I, Geology of the First Geological District* (1843). Note the unconformity between the horizontal limestone above and the steeply dipping strata below.

was intended for use in public schools throughout the state. Both Hall and his close friend Louis Jean Agassiz (a Swiss naturalist noted for his theory of continental glaciation) published scathing critiques, insisting that Foster's chart be legally banned.

In rebuttal, on March 7, 1851, Foster filed lawsuits for libel against both malcontents, thereby feeding fuel to the already intense fire. Further compounding the issue, Foster engaged Emmons as a consultant for a revision of his chart. Naturally, the new version included Emmons' Taconic System with its "Primordial Fossils," which was an anathema to Hall—and to Agassiz. Hall was enraged that this audacious tyro had preempted his own chart for public use. Interestingly, shipment of Foster's revised chart disappeared while in transit aboard a Hudson River steamboat bringing the chart to Albany! A new chart promptly appeared, authored by James Hall and destined for distribution throughout New York State public schools ... and, of course, omitting mention of Emmons' Taconic System.

Seldom have geologists been compelled to defend their ideas in a courtroom. In 1851, Foster's lawsuit came to trial in Albany. Foster essentially represented the ideas of Emmons while his adversary Agassiz represented the ideas of Hall. In Agassiz's corner were the most prestigious geologists of that day. They included: the eminent Professor James Dwight Dana of Yale University; Dr. Edward Hitchcock of Amherst College (where he was both geologist and college president); Sir Charles Lyell of the British Museum; and William W. Mather, principal geologist of the First Geological District, whose comprehensive book (1843) included the Taconic Mountains but not Emmons' Taconic System. This "super-team" of 1850s geologic intelligentsia was opposed, in Foster's corner, by lesser-known personages. These included: Ebenezer Emmons, Sr. (Hall's former teacher); Professor Chester Dewey (Emmons' former teacher); French paleontologist Professor Joachim Barrande of the University of Prague who had recognized "Primordial Fossils" in central Europe (Bohemia); and Professor Jules Marcou of the Geological Survey of France.[22]

Agassiz, Hall, and the "super-team" attacked Emmons' credibility as a geologist in promoting a sequence of more ancient fossiliferous rocks in the Taconic Mountains. Thusly influenced as to Emmons' alleged incompetence and overwhelmed by the professional status of Agassiz, Hall, and their supporters, the jury decided that Foster failed to produce sufficient grounds for suing Agassiz; the correlative case of Foster versus Hall was quickly withdrawn.

Humiliated by the adverse publicity of the trial, shorn of status in the geologic profession of the north, and excommunicated by his former colleagues, Emmons moved south to North Carolina. There, as State Geologist (appointed 1851), his "born-again" Taconic System prospered. Among historians of geology, Ebenezer Emmons, Sr., is remembered as the pioneer proponent of the existence of Early Cambrian (his "Primordial Fossils") fossil-bearing sedimentary rocks in the eastern United States; some of his Taconic System strata are even older (Late Proterozoic–Hadrynian). In addition, Emmons (and Ford, independently) was the first to recognize the significant boundary between the older, greatly deformed strata east of the Hudson River, and the lesser or undeformed strata to the west. This contact became known as "Logan's Line," named at the persuasion of James Hall for his close friend, the celebrated Canadian geologist Sir William Logan. Judging by priority of discovery, this prominent fault should rightfully be termed "Emmons' Line" or "Ford's Line."

In summary, as so often happens in a dispute, Emmons and Hall and their respective supporters were both partly right and partly wrong. Many Taconic rocks are older than the Late Cambrian Potsdam Sandstone, a few are of the same age as those to the west, and some are younger than those to the west. Localized strata at Becraft Mountain (Fig. 148) and Mt. Ida near Hudson are erosional outlier extensions of the Helderberg and Tristates Groups west of the Hudson River. As long as geologists see and interpret data differently, the Taconic Mountains will continue to be the subject of spirited discussion.

Despite the general acceptance of his contributions, the stigma of his court loss together with the prestige of his opposition virtually terminated Emmons' standing in the geologic community. He died on October 1, 1863—a northern sympathizer in the confederacy. Ironically, his son, Ebenezer Emmons, Jr., worked for over 40 years with the very person—James Hall—who was most instrumental in destroying his father's career. Ebenezer Emmons, Jr., died in Albany in 1912.

Ensuing Contributions

Later nineteenth-century geologists who made significant contributions to Taconic geology in Columbia County were T. Nelson Dale of the U. S. Geological Survey and John C. Smock, then a resident of Hudson, New York (his home, now named Cavell House, is part of Columbia Memorial Hospital). Dale (1893, 1904) elaborated on the structure of the Rensselaer Plateau (the upland region of Rensselaer County) and prepared a preliminary colored geologic map between the Hoosic and Kinderhook waterways. Smock (1889) prepared a definitive paper on iron mines and iron ore districts in New York State.

During the twentieth century there emerged an intoxicating fervor to solve the long-standing puzzle of the geologic history and mechanics of origin of the Taconic Mountains. Undeniably, the most incessant and inventive period within this time frame were the halcyon days of 1955–1980. Special plaudits go to the late Emeritus Professor John Rodgers of Yale University, one of the brightest stars amid the constellation of Taconic geologists, for his exhaustive synthesis (1971) of Ordovician structure and strata in the larger Appalachian framework. The author personally benefited from many field trips with Rodgers, not only in the Taconics, but elsewhere in New York, Ontario, Quebec, Vermont, Massachusetts, and Pennsylvania.

To professional and amateur paleontologists, the Taconics are an anathema. The very name connotes unsuccessful fossil collecting because of the deep-sea depositional realms, intense rock deformation, and low-medium grade metamorphism that would have obliterated any fossils had they been there. Thus, fossil-seeking detectives have shunned the region. But the aforementioned obstacles did not deter the extraordinary, multifaceted scientist Dr. Franco Rasetti. Rasetti was a professor of physics at Johns Hopkins University (he was an associate of Drs. Enrico Fermi and Robert Oppenheimer in the development of the atomic bomb). On field trips with this author, however, he demonstrated his proficiency in botany (indeed, he had authored a number of volumes on Alpine wildflowers) and entomology. In addition, it was his passion for Cambrian-age trilobites that prompted him to seek them out in the fossil-poor Taconic strata. His efforts proved eminently successful. In describing several genera and many species, he demonstrated the presence of Medial Cambrian strata in the Taconic Sequence (Rasetti, 1967) for the first time.

Though fossils of the extinct graptolites are regrettably scarce, difficult to identify, and taxonomically puzzling, they are nevertheless superb index fossils. The New York graptolite studies of Ruedemann (1904, 1908, 1942), Berry (1962, 1968), and Riva (1969, 1974) have permitted precise placement within the standard graptolite zonation of the Ordovician Period. This has validated correlation on a global scale. In the Mid-Hudson Valley, graptolites have been found in the shales (usually black) of the Germantown, Stuyvesant Falls, Indian River, Mt. Merino, Austin Glen, and Snake Hill Formations.

To those who map bedrock geology and to stratigraphers (those geologists who are concerned with determining the correct sequence of mappable rock units and properly describing them), the name Taconic conjures up unsettled relationships to other New York formations of comparable age and non-standardized rock unit terminology. The efforts of the following have partly clarified, and in some cases added to, these uncertainties: Bird (1962), Craddock (1957), Elam (1960), Fisher (unpub.), Fisher and Warthin (1976), Goldring (1943), Knopf (1962), Potter (1972), Ratcliffe (1974), Ruedemann (1942), Weaver (1957), Zen (1961), and Zen and Ratcliffe (1971).

To structural geologists, Taconic geology displays an interdependent fabric of faults, folds, gravity slides, thrust slices, cleavage, and regional metamorphism. Differing interpretations persist, especially relating to the distance, or lack of, rock transportation, the mechanics of deformation and the chronology of structural events. In these areas the following are pertinent: Balk (1953), Bird (1969), Bird and Dewey (1975), Bucher (1957), Kidd, Plesch, and Vollmer (1995), Knopf (1962), Ratcliffe (1974), Zen and Hartshorn (1966), and Zen and Ratcliffe (1971).

Notes

1. The popular tourist attraction Crater Lake, Oregon, is not a meteorite impact crater. It is a caldera produced when a large volcano, named Mt. Mazama, erupted explosively approximately 5,000 years ago.
2. Some creatures resembling the plesiosaur have been reported—although little scientifically verifiable evidence has been produced—in several freshwater bodies worldwide, such as at Loch Ness, Scotland and Lake Champlain, New York.
3. Even within the trilobites, ollenellids (Early Cambrian) and phacopids (Early Ordovician–Late Devonian) can be age-distinguished.
4. The name is derived from *Cambria*, the Roman name for Wales, where rocks of this age were first studied by Adam Sedgwick (1785–1873), British stratigrapher and Professor of Geology at Cambridge University.
5. Problematica is the taxonomic name applied to fossils whose classification is unknown or controversial (see D. W. Fisher, *Treatise on Invertebrate Paleontology*, Vol W, pp. W116–W134, 1962)
6. Derived from the ancient tribe of Ordovices, inhabitants of northern Wales where rocks of this age occur. The name Ordovician was a compromise proposal by Charles Lapworth, British paleontologist known for his work on graptolites, which peacefully resolved the controversy between Prof. Adam Sedgewick on the upper limit of the Cambrian and Sir Roderick Murchison on the lower limit of the Silurian.
7. The origin of the name may be traced to D. W. Caldwell. Referring to the geology of Maine, he stated that "The Boil Mountain ophiolite [ultra-mafic scraps of oceanic crust forced onto continental crust] was shoved on to the Chain Lakes massif in Early Ordovician time during the Penobscot mountain building event, which long preceded the Taconic mountain building event." (Caldwell, 1998).
8. This is the "Vermontian Phase" named by the late Professor Marshall Kay of Columbia University, though he did not outline the mechanics of getting the *Taconic Sequence* over the Berkshires.
9. It is interesting to consider the speeds at which tectonic compression may have moved these slices. Using estimates of the distance allochthons moved (perhaps 100 miles), combined with reasonable assumptions of the time involved (perhaps 5 million years), one may conclude that the average rate of movement was on the order of 1 inch/year. Thus, in an average person's lifetime, a slice could move a distance equal to a person's height. This calculation gives a result similar to the measured speeds of diverging plates near mid-ocean ridges today—verifying that our general assumptions about the displacements and time scales involved in slide movements are correct.
10. Zen's Forbes Hill "Conglomerate" is technically a mélange composed of large carbonate blocks of Glens Falls, Isle LaMotte (Fig. 37), Providence Island, Fort Cassin, and Fort Ann limestones in Snake Hill Shale.
11. Mt. Greylock, the highest mountain in Massachusetts, is on the Greylock thrust slice, west of the Berkshire Mountains. This slice may be either the youngest or among the youngest of the Taconian thrust slices—or the phyllite and schist-bearing backside of the Everett Slice.

12. This is the *Hudson River Phase* of the late Emeritus Professor John Rodgers of Yale University, though he never continued its effects into the succeeding Silurian Period.

13. Name derived from the Silures, an ancient people of northern Wales, where the strata were first studied by Roderick Impey Murchison (1792–1871), eminent British geologist and director-general of the Geological Survey of the United Kingdom.

14. Name taken from Devonshire, England, where rocks of this age were first studied by Sir Roderick Murchison.

15. Scottish geologist whose lucid popular writings on geology endeared him to professional geologists and the curious populace alike.

16. Winifred Goldring's superb *Guide to the Geology of John Boyd Thacher Park*, originally published in 1933, was reprinted in 1997 and is highly recommended for those readers interested in "getting out among the rocks."

17. Cameron's Line is a significant structural discontinuity marking the western margin of the Connecticut Valley synclinorium of intensely deformed rocks of Cambrian through Permian ages, metamorphosed by the Taconian, Acadian, and Alleghenyan orogenies.

18. For identification, see *Treatise on Invertebrate Paleontology*, Part H, Brachiopoda (1965).

19. An excellent and educational display of Early Jurassic dinosaur tracks is found in the Connecticut River Valley at Dinosaur State Park, one mile east of exit 23 on Interstate 91 a few miles south of Hartford, Connecticut.

20. Letter of January 12, 1836 to James Hall.

21. See Hernick (1999) for a splendid critical evaluation of the life and contributions of Silas Watson Ford.

22. Marcou never mentioned James Hall in his book *Geology of North America*. Hall considered this a personal affront and never forgave Marcou.

Glossary

allochthonous [ah-lock'-thon-us]—formed elsewhere than in its present place; of foreign origin or introduced by traveling a substantial distance.

amphibolite [am-fib'-oh-lite]—metamorphic rock consisting mainly of amphibole and plagioclase with little or no quartz; formed under moderate to high pressures and temperatures in the 450–700°C range.

anorthosite [an-or'-tho-site]—plutonic igneous rock composed almost entirely of plagioclase feldspar (usually labradorite); forms the "high peaks" (McIntyre range: Mt. Marcy, Whiteface Mountain, etc.) of the Adirondack Mountains.

argillaceous [arj'-ilay-shus]—sediment or sedimentary rock largely composed of clay-sized particles or clay minerals (argillaceous sand, argillaceous limestone).

argillite [arj'-ill-ite]—compact rock derived from either mudstone or shale that has undergone a somewhat higher degree of compaction and lacks the fissility of shale and the cleavage distinctive of slate.

arkose—orange, pink, or red feldspar-rich sandstone; typically coarse-textured and composed of angular or subangular fragments or grains either poorly or moderately sorted. Usually derived from the disintegration of granitic rocks.

autochthonous [aw-tock'-thon-us]—formed in the place where now found.

biotite—"black mica"; potassium-magnesium-iron-fluorine hydroxy silicate.

blastoid [blas'-toyd]—extinct (Ordovician to Permian) class of echinoderms characterized by highly developed five-way radial symmetry; "cousins" to crinoids and cystoids.

brachiopod [brak'-ee-oh-pod]—solitary marine invertebrate characterized by two bilaterally symmetrical but different shells attached along a hinge ("Lamp shell").

breccia [bretch'-ee-ah]—coarse-grained clastic rock composed of angular, broken rock fragments held together by a mineral cement or in a sand or clay matrix. Differs from a conglomerate in that the fragments have sharp edges and unworn corners. Breccias may be igneous, sedimentary, or tectonic in origin.

bryozoan [bry'-oh-zoan]—lacy or fan-shaped invertebrate characterized by colonial growth and a calcareous skeleton.

calcitic dolostone—dolostone containing 50–90% dolomite and 10–50% calcite.

cephalopod [seph'-ah-low-pod]—marine mollusk with straight to variously coiled shell divided into chambers with a connecting canal (siphuncle) allowing exchange of water and air. Ammonites, ceratites, and goniatites are extinct; modern nautiloids are represented by a single-shelled form (*Nautilus*) while octopii and squids lack shells.

charnockite [char-nok-ite]—orthopyroxene granitic plutonic rock formed by high temperature and pressure. Name derived from Job Charnock, founder of Calcutta, India, from whose tombstone the rock was described.

chert—hard, extremely dense or compact, dull to semi-vitreous, sedimentary rock composed of microcrystalline silica (quartz) particles less than 30 microns in diameter. Occurs as nodules or discrete beds in limestones, dolostones, and shales. Shows a splintery to conchoidal fracture and variously colored grays, greens, tans, reds, yellows, and black. (See also **flint**.)

chlorite—iron-magnesium-manganese hydroxy aluminum silicate.

clathrate—a compound formed by the inclusion of molecules of one kind in cavities of the crystal lattice of another.

cleavage (mineral)—the natural breaking of a mineral along its crystallographic planes, thus reflecting its crystal structure.

cleavage (structural geology)—tendency of a rock to split along secondary aligned fractures or other closely spaced planar structures or textures. It is produced by deformation and/or metamorphism.

conodont [koh'-nah-dont]—phosphatic microfossil; toothlike in form but not in function, bilaterally paired, marine fossils of uncertain zoologic affinity; useful for correlation.

crinoid [cry'-noyd]—marine echinoderm characterized by a disk or globular body enclosed by calcareous plates from which feeding "arms" emanate and with a long column ("stem") composed of circular disk-like segments (columnals). The column is attached to the sea floor by a complex holdfast base of branching "rootlike" aggregate.

crossbedding—inclined stratification in which the crossbeds are more than 1 cm in thickness; most common in sandstones and siltstones but rare in limestones.

cross-lamination—inclined stratification in which the cross-laminae are less than 1 cm in thickness; usually within a single bed.

cystoid [sis'-toyd]—extinct (Early Ordovician to Late Devonian) class of echinoderms; "cousins" to blastoids and crinoids.

detritus [dee-tryt'-us]—loose rocks, minerals, and fossils that are worn off or removed by mechanical means as by weathering or erosion.

diamictite [die-ah-mick'-tite]—comprehensive, nongenetic term proposed by Flint, et. al. (1960) for a nonsorted or poorly sorted, noncalcareous, terrigenous sedimentary rock that contains a wide range of particle sizes (boulders to silt) in a muddy matrix (tillite, pebbly mudstone).

diorite [die'-oh-rite]—plutonic igneous rock with 0–5% quartz and about 90% alkali (sodium, potassium) and plagioclase (sodium, calcium) feldspars.

dolomite—the mineral calcium-magnesium carbonate.

dolomitic limestone—a limestone containing 50–90% calcite and 10–50% dolomite.

earthquake—trembling of the Earth caused by abrupt release of slowly accumulated strain. Subduction earthquakes occur along the subducting surface and the lower plate may lock against the upper one to be released at some future time. Deep-slab earthquakes occur where the subducting plate bends and begins to sink. Crustal earthquakes occur along faults in the upper plate.

en echelon—geologic features (e.g., faults) that are in an overlapping or staggered arrangement; each is relatively short but collectively they form a linear zone.

epeirogeny [ep'-eh-raj-ah-nee]—vertical or tilting of the crust affecting broad areas of a continent or ocean, producing plateaus or broad basins.

erratic—pebble to house-size rock carried by glacial or floating ice and deposited at some distance from the outcrop from which it was derived; generally rests on bedrock of different lithology.

esker—narrow, sinuous, steep-sided ridge of stratified sand and gravel deposited by a sub-glacial stream in an ice tunnel or within exposed parallel ice walls of a stagnant or melting glacier.

eurypterid [you-rip'-tur-id]—Extinct arachnid arthropod (Ordovician to Permian) with a segmented body and a chitinous exoskeleton. Some of the world's most exceptional eurypterid fossils are found in New York; the Late Silurian *Eurypterus remipes* is the State Fossil of New York.

feldspar—group of rock-forming aluminum silicate minerals with barium, calcium, iron, potassium, rubidium, sodium, and strontium as replacement elements; group members include the minerals microcline, orthoclase, and plagioclase. Feldspars constitute 60–65% of the Earth's crust; they weather into clay.

flint—archaeological term for **chert** that has been worked by humans into artifacts.

foliation [fo'-lee-ay-shun]—planar arrangement of textural or structural features, especially flattening of constituent grains in metamorphic rocks.

gabbro [gab'-row]—dark green to black, coarse-grained plutonic igneous rock with predominant calcium-plagioclase, hornblende, and pyroxene and little or no quartz; the intrusive equivalent of extrusive basalt.

garnet [gar'-nit]—*The State Gem of New York.* Silicate minerals with varying amounts of aluminum, calcium, chromium, iron, magnesium, manganese, and vanadium. Garnets are relatively hard, brittle, transparent to opaque, resinous to vitreous in luster, and exhibit no cleavage. Commonly found as distinct isometric crystals in metamorphic and, less commonly, in igneous rocks. The Barton Mine on Gore Mountain in the Adirondacks was once the world's largest producer of industrial grade garnet.

gneiss [nice]—foliated rock formed by regional metamorphism in which bands or small lenses of granular minerals alternate with bands or small lenses of flaky elongate minerals.

goethite [go'-thyt]—yellow-brown-black mineral, FeO(OH); common constituent of many forms of natural rust or of limonite.

granite—coarse-grained plutonic igneous rock in which orthoclase or microcline feldspar accounts for 50–90% and quartz 10–50%; mica group minerals and hornblende are accessory minerals.

granulite—coarse-textured metamorphic rock composed of uniform-sized granule minerals including quartz, feldspars, amphiboles, pyroxenes; displays definite banding, a crude gneissic structure, owing to parallelism of flat lenses of quartz and/or feldspar; formed at high temperature and high pressure.

graptolite [grap'-toe-lite]—extinct, marine, probably floating organism; fossils are characterized by carbonaceous, elongated blades with serrated edges. Graptolites are most abundant in black shales and are excellent for correlation of strata on an international basis.

gravity slide—caused by avalanching of a rock mass into sediment due to gravity. Denotes rock sliding over a lubricated mud or rock surface.

"greenstone"—compact, greenish-black, dark green to yellowish-green, metamorphosed basic igneous rock (including basalt, gabbro, and pyroxenite) that owes its color to the minerals chlorite, actinolite, or epidote.

hematite [hem'-ah-tite, hee'-ma-tite]—iron (ferrous) oxide, Fe_2O_3.

Iapetus [eye-ap'-eh-tus]—ancient ocean that preceded the Atlantic Ocean and possessed a different configuration. It was the recipient of the marine sediments during theLate Proterozoic and Early Paleozoic in eastern North America.

imbricate structure—tectonic structure displaying a series of nearly parallel and overlapping thrust faults, high-angle reverse faults, or slides, slices, sheets, or wedges that are all inclined in the same direction—toward the source of stress (shingle-structure).

isograd [eye'-so-grad]—line marking the first appearance of a metamorphic mineral in the regional series of increasing metamorphism (chlorite, muscovite, biotite, garnet, staurolite, kyanite, sillimanite).

isostasy [eye-sos'-ta-see]—adjustment of the crust of the Earth in order to maintain equilibrium among units of varying mass and density.

kame—low mound, knob, or hummock composed of stratified sand and gravel deposited by a glacial stream as a fan or delta at the margin of a melting glacier or at a low place or hole on the glacier's surface.

karst—type of topography formed on limestone, dolostone, or other relatively soluble rocks by solution; characterized by sinkholes, caves, and underground drainage.

kettle—steep-sided basin or bowl-shaped hole, usually with interior drainage, in glacial drift deposits. Formed by the melting of a large, detached block of ice within the drift—a collapse feature.

klippe (pl. klippen)—an isolated rock body that is an erosional remnant of a nappe.

light-year—light travels at 186,000 miles/second; a light-year is the distance light travels in one year—approximately 5,700,000,000,000 miles.

limonite [lye'-mah-night]—general term for a group of dark brown, yellowish-brown, yellow, red, or black hydrous iron oxides whose identities are unknown. Minor ore of iron.

magnetite—naturally magnetic mineral; ferric iron oxide, Fe_3O_4 (lodestone).

megabreccia [mega-bretch'-ya]—term used by Landes (1945) for a rock produced by brecciation (crushing) on a very large scale and containing randomly oriented blocks. Longwell (1951) used the term for a coarse breccia containing blocks as large as 400 meters long and developed downslope from large thrust faults by gravitational sliding. It is partly tectonic and partly sedimentary in origin.

metamorphism [met'-ah-more-fizem]—chemical, mineralogical, and structural adjustment of rocks to physical and chemical conditions (typically heat and/or pressure) which differ from the conditions under which the rock originated.

mine—underground excavation for the extraction of economically useful mineral deposits or building stone.

moraine [mah-rain']—mound, ridge, or other accumulation of unsorted, unstratified glacial drift (usually till) deposited by glacial ice.

olistostrome [oh-liss'-toe-strohm]—from the Greek *olistomai* (to slide) and *stroma* (bed), originates by submarine gravity sliding as a sedimentary deposit.

orogeny [o-raj'-ah-nee]—process by which structures were formed within folded mountains; includes thrust-faulting in the outer and higher layers and plastic folding, metamorphism, and igneous intrusion in the inner and deeper layers; implies compression of the crust.

orthoclase [or'-tho-clayz]—potassium-sodium-aluminum silicate; light colored feldspar (white, tan, pink, light gray); common in "acid"-type igneous rocks (granites, rhyolites, diorites, trachytes).

ostracode [os'-tra-cod]—bean-shaped, bivalve marine and freshwater crustaceans with seven pairs of appendages.

parautochthonous [pear-aw-tock'-thon-us]—rocks that are transported but still within their depositional realm—not far-traveled.

phyllite [fill'-ite]—metamorphic rock intermediate in grade between slate and schist; minute crystals of chlorite and sericite impart a silky sheen to the cleavage surfaces, which may be corrugated.

pit—surface working for the extraction of clay, sand, and/or gravel.

plagioclase [plaj'-ee-oh-clayz]—green, bluish-gray to dark gray "alkali" feldspar; calcium-aluminum silicate; the common mineral in "basic" igneous rocks (anorthosites, gabbros, basalts, hornblendites, pyroxenites).

plate tectonics—study of crustal plate movements, interactions, and their construction and destruction; the method of explaining world seismicity, volcanism, mountain building, and paleomagnetism in terms of plate motions.

plutonic—crystalline rock formed by the solidification of magma deep within the Earth.

pothole—circular or elliptical hole in a rock bed of a stream, formed by the grinding action of stones and coarse sediment whirled around by moving current or falling water.

prospect—area that is a potential site of mineral deposits, but is not currently producing or has not given evidence of economic value.

pyroxenite [peer-ox'-in-ite]—ultramafic plutonic rock chiefly composed of pyroxene, with accessory hornblende, biotite, or olivine.

quadrangle—a four-sided map such as published by the U.S. Geological Survey.

quarry—surface working for the extraction of stone.

quartz—silicon dioxide, SiO_2; may be cryptocrystalline (chalcedony, chert, jasper, and agate) or in hexagonal-shaped crystals with pyramid and prism faces. "Herkimer diamonds" are doubly terminated quartz crystals.

quartzite (meta)—metamorphic rock consisting mainly of quartz and formed by regional or contact metamorphism of sandstone, siltstone, or chert.

quartzite (sed)—very hard unmetamorphosed sandstone in which the quartz grains have been cemented with secondary silica (quartz) so that the rock breaks across the grains rather than around them (orthoquartzite).

relict—a physical feature, mineral, structure, etc., remaining after other components have wasted away or been altered; a plant or animal species living on in isolation in a small local area as a survival from an earlier period or as a remnant of an almost extinct group.

rhyolite [rye'-oh-lite]— light-colored igneous rock, the fine-grained equivalent of granite; commonly porphyritic; quartz and orthoclase dominate.

rift—long, narrow, continental trough bounded by normal, high-angle faults. It is a portion of the crust ruptured by extension (tension) during a taphrogeny. Also called a graben [grah'-bin].

schist—metamorphic crystalline rock exhibiting closely foliated structure and which can be split along parallel planes.

schistosity [shis-tos'-eh-tee]—foliation in schist, phyllite, or other coarse-grained crystalline rock due to the parallel arangement of mineral grains of the platy, prismatic, or ellipsoidal types.

siderite [syd'-ur-ite]—iron carbonate, $FeCO_3$; commonly yellow-brown, brown-red, or brown-black; valuable ore of iron.

slate—compact, fine-grained metamorphic rock, formed from shale or argillite; can be split into slabs and thin plates making it economically useful for patio and sidewalk flagging and roofing.

stratigraphy [strah-tig'-rah-fee]—science of description, arrangement, classification, and correlation of sedimentary rocks; may include the interpretation of ancient sedimentary environments (paleoecology).

stratum [strat'-um, straight'-um]—tabular or sheet-like layer of sediment or sedimentary rock.

stromatolite—laminated alternation of sediment (usually calcium carbonate) and blue green algae (cyanobacteria).

stromatoporoid—extinct group of reef-building marine organisms consisting of laminated and nodular masses; now thought to have affinities with the sponges.

syenite—coarse-grained plutonic igneous rock somewhat more mafic than granite with small amounts of plagioclase feldspar and hornblende.

talus (syn. scree)—rock fragments and sediments of any size and shape derived from and lying at the base of a cliff or rocky slope.

taphrogeny [taf-raj'-ah-nee]—regional development of downdropped (rifts, graben) and uplifted (horsts) segments of the Earth's crust by high-angle normal faulting; denotes tension of the crust; opposite of orogeny, which denotes compression of the crust.

tectonics [tek-taan'-iks]—branch of geology dealing with the architecture of the outer part of the Earth. It is the regional assembling of deformational and structural features, a study of their mutual relations, origin, and historic evolution.

tektites—glassy beads, from microscopic in size to several inches across, found in aerodynamic shapes; they were most likely created when meteorite and asteroid impacts ejected molten droplets of material at supersonic speeds into the atmosphere, where they solidified and fell back to Earth.

terrane (terrain)—area or surface over which a particular rock or sediment, or group of rocks or sediments, is prevalent; the physical features of a tract of land.

till—heterogeneous mixture of unsorted and unstratified boulders, gravel, sand, silt, and clay deposited directly by, and underneath, a glacier without reworking by meltwater; tillite is a consolidated till.

tonalite [tone'-ah-lite]—plutonic igneous rock with 10–60% quartz and the remainder essentially of one or more alkali (sodium, potassium) and plagioclase (sodium, calcium) feldspars, i.e., a quartz-rich diorite.

trench—narrow, elongate depression of the sea floor, with steep sides, typically oriented parallel to the trend of the continent and existing between the continental margin and the abyssal hills. Trenches are much deeper than the surrounding ocean floor and may be thousands of kilometers long. Trenches usually form at convergent plate boundaries.

trilobite [tri'-low-bite]—extinct marine crustacean characterized by a three-lobed, ovoid to sub-elliptical exoskeleton divisible longitudinally into axial and flanking side regions and transversely into head (cephalon), body (thorax), and tail (pygidium).

tuff—general term for all volcanic (pyroclastic) rocks; some tuffs tend to be stratified and appear as sedimentary rocks.

turbidite [tur'-bid-ite]—sediment or rock deposited by a turbidity current; characterized by graded bedding, moderate sorting, and well-developed primary sedimentary structures (lode casts, rill marks, scour troughs, and complex ripple marks).

turbidity current [tur-bid'-eh-tee]—density current in water brought about by different amounts of matter in suspension. A bottom-flowing current is laden with suspended sediments and, under gravity, moves swiftly downslope distributing the sediments horizontally on the sea floor in a wide variety of sedimentary patterns.

ultramafic—term for an igneous rock composed of extremely basic minerals (rich in iron and magnesium) including augite, hornblende, hypersthene, olivine, and pyroxene.

wildflysch [vild'-flish]—mappable stratigraphic unit displaying various types and sizes of irregularly sorted rocks resulting from tectonic fragmentation and slumping or sliding under the influence of gravity.

zone—minor interval in any category of classification; examples are biozone, lithozone, chronozone, mineral zone, metamorphic zone, etc. Term should always be preceded by a modifier indicating the kind of zone in which reference is made.

Selected Bibliography

"Some books are to be tasted, others to be swallowed, and some few to be chewed and digested."

Francis Bacon

Not all of the references used in the preparation of this book are listed here. Most published from the nineteenth through the first half of the twentieth century are out of print and will be difficult to locate. Many of the more recent references used to complete this work are written for the professional geologist. While possibly available at college and university libraries, they contain obscure and technical terminology and will be of limited use to the lay reader. Should you seek additional information on the topics discussed, the following references can be more readily located. Data used in the preparation of the Tectonic Map of Columbia County (Plate 1) are cited on that sheet.

The Internet as a Resource

The importance and usefulness of the Internet to those who would like to stay up-to-date with the latest research in our increasingly fast-paced world cannot be overstated. Most states with geological survey offices maintain Web sites, and most college geology departments have online information concerning their research. As an increasing number of universities and government agencies convert older (and essentially unavailable) documents to an electronic format, these important resources become more easily available to all of us. As an example, in 1961 the *Centennial Geologic Map of Vermont* was published by the Vermont Geological Survey. Today this map is out of print; however, by going to: http://www.anr.state.vt.us/dec/geo/vgs.htm, this map may be downloaded and printed for detailed study.

Beck, Lewis Caleb. (1842) *Mineralogy of New York.* Albany, N.Y. (536 pages, 7 plates)

Berry, William B. N. (1962) "Stratigraphy, zonation, and age of Schaghticoke, Deepkill, and Normanskill Shales, eastern New York." *Geological Society of America Bulletin,* 73: 695–718. (2 plates, 3 figures)

———. (1963) "Ordovician correlations in the Taconic and adjacent regions. Guidebook for field trips, 1963 meeting." *Geological Society of America Bulletin.*

Bird, John M., and Franco Rasetti. (1968) "Lower, Middle, and Upper Cambrian faunas in the Taconic Sequence of eastern New York: stratigraphic and biostratigraphic significance." *Geological Society of America Special Paper 113.* (66 pages, 14 figures)

Bird, John M., and John F. Dewey. (1975) "Selected localities in the Taconics and their implications for the plate tectonic origin of the Taconic region." *Guidebook for Field Trips, New England Intercollegiate Geological Conference:* 87–121.

Briggs, Derek E. G., Douglas H. Erwin, and Douglas H. Collier. (1994) *The Fossils of the Burgess Shale.* Washington and London: Smithsonian Institution Press. (238 pages, 180 photos and drawings, 24 figures)

Cadwell, Donald H. (1986) "Late Wisconsinan Stratigraphy of the Catskill Mountains in the Wisconsinan Stage of the First Geological District, Eastern New York." *New York State Museum Bulletin 455,* 73–88. (7 figures, 2 tables)

Caldwell, Dabney W. (1998) *Roadside Geology of Maine.* Missoula, Mont.: Mountain Press Publishing. (317 pages, profusely illustrated)

Clarke, John Mason. (1900) "The Oriskany Fauna of Becraft Mountain, Columbia County, N. Y." *New York State Museum Memoir 3.* (128 pages, 9 plates)

Connally, G. Gordon, and Les A. Sirkin. (1973) "Wisconsinan history of the Hudson–Champlain Lobe," in Black, R. F., R. P. Goldthwait, and H. B. Willman, eds., "The Wisconsinan Stage," *Geological Society of America Memoir 136*, 47–69.

———. (1986) "Woodfordian Ice Margins, Recessional Events, and Pollen Stratigraphy of the Mid-Hudson Valley," in Cadwell, Donald H., ed., *The Wisconsinan Stage of the First Geological District, Eastern New York,* 50–72. (9 figures, 4 tables)

Dale, T. Nelson. (1904) "Geology of the Hudson Valley between the Hoosic and the Kinderhook." *U. S. Geological Survey Bulletin 242.* (63 pages, colored geological map)

Doll, Charles G., et al. (1961) *Centennial Geologic Map of Vermont.* Vermont Geological Survey. (1:250,000, color)

Dunn, James R., and Donald W. Fisher. (1954) "Occurences, properties, and paragenesis of anthraxolite in the Mohawk Valley, New York." *American Journal of Science,* 252, 489–501.

Emmons, Ebenezer. (1846) "The New York System." *Agriculture of New York,* 1, 113–206.

———. (1846) "The Taconic System." *Agriculture of New York,* 1, 45–112.

———. (1849) "On the Identity of *Atops trilineatus* and *Triarthrus beckeii* (Green) with remarks upon the *Elliptocephala asaphoides." American Association for the Advancement of Science, Proceedings,* 1, 16–19.

———. (1855) "The Taconic System." *American Geologist,* 1 (2). (251 pages)

Fisher, Donald W. (1957) "Mohawkian (Middle Ordovician) Biostratigraphy of the Wells Outlier, Hamilton County, New York." *New York State Museum Bulletin 359.* (33 pages, 6 plates, 2 figures, 1 photo)

———. (1961) "Stratigraphy and structure in the southern Taconics (Rensselaer and Columbia Counties, New York)." *Guidebook for Field Trips, New York State Geological Association 33rd Annual Meeting,* Troy, New York, D1–D27. (4 plates)

———. (1968) "Geology of the Plattsburgh and Rouses Point, New York–Vermont Quadrangles." *New York State Museum Map and Chart Series 10.* (51 pages, 37 figures, Plate 1 [colored geologic map, 1:62,500], Plate 2 [5 colored cross sections])

———. (1977) "Correlation of the Hadrynian, Cambrian, and Ordovician Rocks in New York State." *New York State Museum Map and Chart Series 25.* (75 pages, 76 figures, 5 plates [4 in., color])

———. (1978) "Amos Eaton—Passionate Peddler of Science." *The Conservationist,* 32, no. 4, 36–39. (5 illustrations)

———. (1978) "Laudable Legacy—A synopsis of the titans of geology and paleontology in New York State." *Guidebook for Field Trips, New York Geological Association Guidebook 50th Annual Meeting,* Syracuse University, 1–24. (9 plates)

———. (1979) "Folding in the foreland, Middle Ordovician Dolgeville facies, Mohawk Valley, New York." *Geology,* 7, 455–459. (7 figures)

———. (1980) "Bedrock Geology of the Central Mohawk Valley, New York." *New York State Museum Map and Chart Series 33,.* (44 pages, 64 figures, 2 colored maps: 1:48,000)

———. (1981) "Emmons, Hall, Mather, and Vanuxem—the four 'horsemen' of the New York State Geological Survey." *Northeastern Geology,* 3, 29–46.

———. (1984) "Bedrock Geology of the Glens Falls–Whitehall Region, New York." *New York State Museum Map and Chart Series 35.* (58 pages, 90 figures, Plate 1 [colored map, 1:48,000], Plate 2 [structure cross-sections], Plate 3 [stratigraphic profiles and paleogeographic maps])

———. (1991) "Algae from Antiquity." *The Conservationist,* 45, no. 4 (1991), 42–47. (8 illustrations)

Fisher, Donald W., Yngvar W. Isachsen, and Lawrence V. Rickard. (1962) *Geologic Map of New York. New York State Museum Map and Chart Series 15;* reprinted, 1970, 1995. (5 colored sheets, 1:250,000)

Fisher, Donald W., and A. Scott Warthin. (1968–1980) *Bedrock geology of the Mid-Hudson Valley.* Unpublished, New York State Geological Survey, Albany, New York. (1:24,000 quadrangles [Hopewell Junction, Hyde Park, Kingston East, Millbrook, Newburgh, Pine Plains, Pleasant

Valley, Poughkeepsie, Rock City, Salt Point, Wappingers Falls])

Fisher, Donald W., and A. Scott Warthin. (1976) "Stratigraphy and structural geology in western Dutchess County, New York." *Guidebook for Field Trips, New York State Geological Association 48th Annual Meeting,* Vassar College, Poughkeepsie, B6-1–B6-36. (13 figures, black and white geologic map)

Ford, Silas W. (1884) "Note on the discovery of primordial fossils in the town of Stuyvesant, Columbia County, New York." *American Journal of Science,* 3 (28), 35–37.

Ford, Silas W., and William B. Dwight. (1886) "On fossils from metamorphic limestones of the Taconic series of Emmons at Canaan, N. Y." *American Journal of Science,* 3 (31), 248–255. (illustrations)

Freidman, Gerald M. (1956) "The origin of the spinel–emery deposits with particular reference to those of the Cortlandt Complex of New York." *New York State Museum Bulletin 251.* (68 pages)

Funk, Robert, Donald W. Fisher, and Edgar M. Reilly, Jr. (1970) "Caribou and Paleo-Indians in New York State: a presumed association." *American Journal of Science,* 268, 181–186. (1 figure, 1 plate)

Gibson, G. G., S. A.Teeter, and M. A. Fedonkin. (1984) "Ediacaran fossils from the Carolina Slate belt, Stanley County, North Carolina." *Geology,* 12, 387–390.

Goldring, Winifred. (1938) "Algal Barrier Reefs in the Lower Ozarkian of New York." *New York State Museum Bulletin 315.* (75 pages, 22 figures)

———. (1997, special reprint) "Guide to the Geology of the John Boyd Thacher State Park (Indian Ladder Region) and Vicinity." *New York State Museum Handbook 14,* E. Landing and J. B. Skiba, eds. (112 pages, 32 figures, maps)

Gould, Stephen Jay. (1989) *Wonderful Life.* W. W. Norton and Company. (347 pages, 116 figures)

Grabau, Amadeus W. (1903) "Stratigraphy of Becraft Mountain, Columbia County, New York." *New York State Museum Bulletin 69,* 1030–1079. (13 figures [colored geologic map, 1:10,560])

Hernick, Linda V. (1999) "Silas Watson Ford: a major but little-known contributor to the Cambrian Paleontology of North America." *Earth Science History,* 18, no. 2, 246–263. (4 figures)

———. (2003) "The Gilboa Fossils." *New York State Museum Circular 65.* (100 pages, 47 figures)

Isachsen, Yngvar W., S. W. Wright, and Frank A. Ravetta. (1994) "The Panther Mountain Circular Feature Possibly Hides a Buried Impact Crater." *Northeastern Geology,* 16, no. 2, 123–136. (7 figures)

Kidd, William S. F., Andreas Plesch, and Frederick W. Vollmer. (1995) "Lithofacies and structure of the Taconic Flysch, Mélange, and Allochthon in the New York Capital District." *Guidebook for Field Trips, New York State Geological Association 67th Annual Meeting,* Union College, Schenectady, New York, 57–80. (11 figures)

Knopf, Eleanora Bliss. (1962) "Stratigraphy and structure of the Stissing Mountain area, Dutchess County, N. Y." *Stanford University Publications in the Geological Sciences,* 7, no. 1. (55 pages, colored geologic map [1:24,000])

Marshak, Stephen. (1990) "Structural Geology of Silurian and Devonian strata in the Mid-Hudson Valley, New York." *New York State Museum Map and Chart Series 41.* (66 pages, 48 figures, 3 plates)

Mather, William Williams. (1843) *Geology of New York, Part 1, Comprising the First Geological District.* Albany, New York. (653 pages, 35 figures, 45 plates [includes colored maps and cross sections])

McAllister, Ethel M. (1941) *Amos Eaton, Scientist and Educator, 1776–1842.* Philadelphia: University of Pennsylvania Press. (587 pages, 8 illustrations)

McLelland, James M., and Donald W. Fisher. (1976) "Stratigraphy and structural geology in the Harlem Valley, southeast Dutchess County, New York." *Guidebook for Field Trips, New York State Geological Association 48th Annual Meeting,* Vassar College, C7-1–C7-37. (6 figures)

Offield, Terry W. (1967) "Bedrock Geology of the Goshen-Greenwood Lake Area, New York." *New York State Museum Map and Chart Series 9.* (78 pages, 56 figures, 1:48,000 colored geological map)

Potter, Donald B. (1972) "Stratigraphy and Structure of the Hoosick Falls Area, New York–

Vermont, East–Central Taconics." *New York State Museum Map and Chart Series 19.* (71 pages, 42 figures, 2 plates [colored geologic map; 1:24,000, cross sections])

Rasetti, Franco. (1966) "New Lower Cambrian trilobite faunule from the Taconic Sequence of New York." *Smithsonian Miscellaneous Collections,* 148 (9). (52 pages, 12 plates)

———. (1967) "Lower and Middle Cambrian trilobite faunas from the Taconic Sequence of New York." *Smithsonian Miscellaneous Collections,* 152 (4). (111 pages, 14 plates)

Rickard, Lawrence V. (1962) "Late Cayugan (Upper Silurian) and Helderbergian (Lower Devonian) Stratigraphy in New York." *New York State Museum Bulletin 386.* (157 pages, 28 figures)

———. (1975) "Correlation of the Silurian and Devonian rocks in New York State." *New York State Museum Map and Chart Series 24.* (text, 4 plates [3 in color])

Riva, John. (1969) "Middle and Upper Ordovician graptolite fauna of the St. Lawrence Lowlands of Quebec and Anticosti Island." *American Association of Petroleum Geologists Memoir 112,* 513–556.

———. (1974) "A revision of some Ordovician graptolites of eastern North America." *Paleontographica 17,* 1–40. (2 plates)

Rodgers, John. (1985) *Bedrock Geological Map of Connecticut.* Connecticut Natural Resources Atlas Series, Connecticut Geological and Natural History Survey in cooperation with the U. S. Geological Survey. (1:125,000, 2 sheets in color, colored cross-sections)

Ruedemann, Rudolf. (1901) "Trenton conglomerate of Rysedorph Hill and its fauna." *New York State Museum Bulletin 49,* 2–114.

———. (1930) "Geology of the Capital District (Albany, Cohoes, Troy, and Schenectady Quadrangles)." *New York State Museum Bulletin 285.* (218 pages, 79 figures, colored geologic map [1:62,500])

———. (1931) "Age and origin of the siderite and limonite of the Burden iron mines near Hudson, New York." *New York State Museum Bulletin 286,* 135–152 (figures 26–28).

Shaw, Frederick C. (1968) "Early Middle Ordovician Chazy trilobites of New York." *New York State Museum Memoir 17.* (163 pages, 24 plates, 16 figures, 8 tables)

Skehan, James W. (2001) *Roadside Geology of Massachusetts.* Missoula, Mont.: Mountain Press Publishing. (379 pages, profusely illustrated)

Van Diver, Bradford. (1987) *Roadside Geology of Vermont and New Hampshire.* Missoula, Mont.: Mountain Press Publishing. (230 pages, profusely illustrated)

———. (1990) *Roadside Geology of Pennsylvania.* Missoula, Mont.: Mountain Press Publishing. (352 pages, profusely illustrated)

———. (2001) *Roadside Geology of New York.* Missoula, Mont.: Mountain Press Publishing. (415 pages, profusely illustrated)

Yochelson, Ellis L. (1998) *Charles Doolittle Walcott, Paleontologist.* Kent State University Press. (520 pages)

Zen, E-an. (1961) "Stratigraphy and structure of the north end of the Taconic Range in west–central Vermont." *Geological Society of America Bulletin,* 72, 293–378. (colored geological map)

———. (1967) "Time and space relations of the Taconic allochthon and autochthon." *Geological Society of America Special Paper 97.* (107 pages)

Zen, E-an, ed. (1983) *Geologic map of Massachusetts.* Compiled by R. Goldsmith, N. M. Ratcliffe, P. Robinson, R. S. Stanley, U. S. Geological Survey. (colored geologic map,
3 sheets 1: 250,000)

Sources of Bedrock Geology for the Tectonic Map of Columbia County, New York

The geological map of tectonic slices in Columbia County presented in this work is a synthesis of new and old. Over the years different workers with varying degrees of geologic expertise have mapped the quadrangles contained within Columbia County to various degrees of completeness and accuracy. In performing his own fieldwork over a fifty-year period, the author has been able to locate and identify rock exposures that were perhaps unavailable to, unnoticed, or misidentified by previous workers, allowing the author to develop this more comprehensive overview of Columbia County's geology.

The following list of references has been annotated to indicate the degree of revision employed in the incorporation of each work:

no no revision; the work has been incorporated essentially unchanged

min minor revision; approximately 10–20% of map area revised

maj major revision; approximately 50–75% of map area revised

Bird, John M. (1962) "Geology of the Nassau Quadrangle." Unpublished Ph.D. thesis, Rensselaer Polytechnic Institute, Troy, N.Y. (204 pages; colored map, 1:24,000) **[no]**

Craddock, J. Campbell. (1957) "Stratigraphy and structure of the Kinderhook Quadrangle, New York and the 'Taconic Klippe.'" *Geological Society of America Bulletin*, 68, 675–724. (colored map, 1:62,500) **[maj]**

Fisher, Donald W. (1955–1981) Unpublished bedrock geology maps (1:24,000) of all quadrangles of Columbia County and western Dutchess County, on file with the New York State Geological Survey, Albany, N.Y.

Goldring, Winifred. (1943) "Geology of the Coxsackie Quadrangle, New York." *New York State Museum Bulletin 322*. (374 pages, 71 figs.; colored map, 1:62,500) **[maj]**

Ratcliffe, Nicholas M. (1974) "Bedrock geologic map of the State Line Quadrangle, Columbia County, New York and Berkshire County, Massachusetts." *United States Geological Survey*, GQ-1142. (colored map, 1:24,000) **[min]**

Ratcliffe, Nicholas M., and Beshid Bahrami. (1976) "The Chatham Fault: a reinterpretation of the contact relationships between the Giddings Brook and Chatham Slices of the Taconic Allochthon in N.Y. State." *Geology*, 4, 56–60. (3 figures) **[maj]**

Ratcliffe, Nicholas M. (1978) "Bedrock geologic map of the Pittsfield West Quadrangle and part of the Canaan Quadrangle, Berkshire County, Massachusetts and Columbia County, New York." *United State Geological Survey Miscellaneous Field Studies Map* MF–980. (colored map, 1:24,000) **[min]**

Rodgers, John. (1975) Unpublished maps (1:24,000) of bedrock geology of parts of the Ancram, Copake, Millerton, and Pine Plains Quadrangles, on file with the New York State Geological Survey, Albany, N.Y. **[no]**

Ruedemann, Rudolph. (1942) "Geology of the Catskill and Kaaterskill Quadrangles. Part 1, Cambrian and Ordovician Geology of the Catskill Quadrangle." *New York State Museum Bulletin* 331. (188 pages, 82 figs.; colored map, 1:62,500) **[maj]**

Weaver, John Dodsworth. (1957) "Stratigraphy and structure of the Copake Quadrangle, New York." *Geological Society of America Bulletin* 68, 725–762, (3 pls., 8 fig.; colored map, 1:62,500) **[maj]**

Zen, E-an, and J. H. Hartshorn. (1966) "Geologic Map of the Bashbish Falls Quadrangle, Massachusetts, Connecticut, and New York." *United States Geological Survey Map* GQ–507. (7 pages, explanatory text; colored map, 1:24,000) **[min]**

Zen, E-an, and Nicholas M. Ratcliffe. (1971) "Bedrock geologic map of the Egremont Quadrangle and adjacent areas, Berkshire County, Massachusetts and Columbia County, New York." *United States Geological Survey, Miscellaneous Geological Investigations Map* I–628. (colored map, 1:24,000) **[min]**

Index

Page numbers in italics refer to figures/illustrations

Index of Selected Fossils

About the Author

The author in a typical field geologist position—nose and eyes on the rock! Photograph by Betty Fisher.

Donald W. Fisher studied at the University of Buffalo where he received a B.A. (1944) and M.A. (1948) in geology. He received his Ph.D. at the University of Rochester in 1952. During his graduate work Fisher taught at Union College in 1949–50 and 1951–52. Later in life, from 1987–92, he returned to teaching and was an instructor at Columbia–Greene Community College. In 1952 he was employed by the New York State Geological Survey. His first position there was Senior Paleontologist; he rose to the position of State Paleontologist in 1955.

During his tenure as State Paleontologist, Fisher's accomplishments included: development of correlation charts for the Cambrian, Ordovician, and Silurian Systems of the State (New York Map and Chart Series 1–3); compilation of a statewide geologic map (with colleagues Ingvar Isachsen and Lawrence Rickard) published in 1962 (Geologic Map of New York); geologic mapping of the Plattsburgh–Rouses Point Quadrangles, the central Mohawk Valley, and the Glens Falls–Whitehall region (New York State Museum Map and Chart series 10, 13, and 35, respectively); organization of field trips for the New York State Geological Association, the New England Intercollegiate Geological Conference, and the Society of Economic Paleontologists and Mineralogists; co-author of *Colossal Cataract: The Geological History of Niagara Falls* (1981): publication of over 100 papers in professional journals; service on graduate student thesis committees at Rensselaer Polytechnic Institute, Syracuse University, Columbia University, and the University of Rochester; and working to promote the successful adoption in 1984 of *Eurypterus remipes* as the New York State Fossil.

In 1994, the New York State Museum honored Fisher by publishing Bulletin Number 481, *Studies in Stratigraphy and Paleontology in Honor of Donald W. Fisher*. Among Fisher's other honors and awards are: Phi Beta Kappa (1948); Sigma Xi (1950); Ralph Digman Award, National Association of Geology Teachers (1983); Distinguished Alumni Award, University of Buffalo (1995); John Mason Clarke Distinguished Service Award, New York State Geological Survey, New York State Museum (1995).

Dr. Fisher lives in Columbia County, where he operates Fisher's O.K. Rock Shop on 2 Chatham Street (U.S. 9) in Kinderhook (518-758-7657).